Backyard Chickens

A Comprehensive Guide to Raising Chickens for Beginners, Including Tips on Choosing a Breed and Building the Coop

Contents

Introduction

Self-sufficiency, especially when it comes to your food sources, can be extremely liberating. No one wants to spend hours reading labels trying to figure out whether or not something is healthy. While you may not be able to control where all your food comes from, raising backyard chickens is one way to make sure that you are getting healthy and wholesome eggs from your own flock.

Chickens are pretty low maintenance, and this is probably the reason raising backyard chickens has become a popular hobby. After all, who would not want a pet that also provides fresh eggs? However, it is not all about the eggs. Chickens are not just a source of eggs, but they also make interesting family pets that your whole family can enjoy.

While you may have an interest in raising chickens in your backyard, you probably do not know where to start or even how to do it. That is why we have compiled a simple and straightforward guide for beginners who want to raise chickens. Some people may have no experience at all with raising chickens, but that should not deter you from taking up this rewarding hobby.

This book gives you detailed information on how to start your backyard flock. We want you to be able to take care of the chickens once you have them, so this book also gives you in-depth insights into how to raise, care for, and maintain your backyard chickens. From building a coop to equipping it will all the accessories you need to take proper care of your chickens, you will get all the information you need to get started.

A significant part of the book is dedicated to the proper care of chicks and how to raise them from day-old baby chicks to adult hens that lay eggs. We understand that caring for pets is a delicate process, and this book aims to equip you with all the knowledge you need to raise a healthy and happy flock.

We have a section devoted to understanding chicken behavior that is geared to help you bond better with your flock. That part of the book is designed to help you pick up on any distress signals in your flock or any signs of illness. If you have thought about raising backyard chickens for a while, this is the book that will guide you on how to do it. You need not be a farmer in a rural setup to raise healthy chickens. With the right tools and information, you can transform your backyard into a safe haven for chickens.

Naturally, the first place to start on your journey to raising backyard chickens is understanding why you need to do it at all.

Chapter 1: Why Raise Chickens at Home?

Raising chickens is no longer just the preserve of people with rural farms. An increasing number of people are turning to raising chickens in their back yards. You have probably thought about it yourself, but, like any other venture, you may still wonder whether the pros outweigh the cons. If the allure of fresh eggs is not enough to convince you, there are still plenty of reasons to raise chickens in your backyard.

Whether you are looking for a fulfilling hobby or, like many others, just want to have more control over what's on your dining table, chickens are a great place to start. For most people who consider raising pets or livestock of any kind, space is usually one of the main deterrents. The beauty in choosing chickens is that they do not take up a lot of room. If you have a small backyard, you can still comfortably accommodate a small flock of chickens. For a modest flock of about six chickens, you would need approximately 110 square feet of space. This is why more and more people have taken up raising chickens in their backyards. With just a modest amount of space, you can have your own supply of fresh eggs readily available and a rewarding hobby to boot.

Another key concern that you may have when deciding to raise chickens is how much time and maintenance is required. Pets come with their fair share of maintenance, and chickens are no different. However, chickens are pretty low maintenance, and this is not one of those pursuits that will take up hours of your time. Keeping chickens will require some effort on your part, but for the most part, these birds are pretty self-sufficient and do not require round-the-clock care.

As long as you have a secure coop to keep your flock safe from predators, you will find that chickens require little attention. Most people find that the daily care required for chickens takes up less than half an hour a day, so this is not one of those hobbies that will turn out to be time-consuming.

If you have a dog or a cat, you will realize that chickens are easier to care for than either dogs or cats. Chickens do not need as much human attention as other household pets. Provided they are well fed and housed, you will find that you can pretty much go about your business without needing to keep checking on your chickens all the time. Most people who keep chickens will tell you that they are easy to keep because they are pretty independent.

For people with children at home, getting pets that your children can be around safely and also help care for is always a plus. As far as pets go, chickens are mostly non-territorial, and most are not aggressive. This means they make great pets that your children can also enjoy taking care of. Nothing teaches young children about responsibility better than having them help out in caring for a pet.

Chicken-watching can also make for a fun and interesting pastime. Chickens have individual personalities and quirks that make them fun to watch. They can also be beautiful, and depending on the breed, some can be unique in terms of appearance. Some chickens are also friendly and will approach you or your kids whenever you are within their sight. This means this rewarding hobby will be anything but dull.

If your backyard is well enclosed, you can let the chickens roam free during the daytime since chickens tend to be pretty friendly and will not be aggressive. Of course, if you are to let your chickens roam free in the backyard, you need to ensure that they are safe from predators.

While having a pet is great, chickens come with the additional benefit of being a source of fresh food in terms of eggs and meat. While you can buy eggs at the grocery store, you can never be sure about the freshness or quality of store-bought eggs. The increasing popularity of raising backyard chickens, even amongst celebrities such as Jennifer Aniston, Martha Stewart, and many more, is partly because people are getting more conscious about the kind of food they eat. When you get your eggs from your own chickens, you know what they have been eating, and you are, therefore, in control of what you are eating. This means you can choose to feed your chickens organic food and, as a result, enjoy organic eggs that are nutritious and free of GMO additives.

When you buy products from a grocery store, you have little knowledge of what kind of produce you are getting or how the chickens that produced the eggs were raised. This means that you cannot be 100% sure about the quality or freshness of the eggs you are getting. But with your own flock, you have control over what they eat, and you gather up the eggs daily, so freshness and quality are guaranteed. This also goes for people who want to keep chickens for meat. With your own chickens, you are assured of quality and that what ends up on your dining table is free of any harmful chemicals and additives.

Research studies show that eggs from free-range chickens tend to have higher concentrations of nutrients such as beta-carotene, omega three, and vitamins A and E than eggs sourced from battery hens or chickens that are raised in cages. So, if you have been on the fence about raising your own chickens, consider the additional health benefits for you and your family that come with having control over the kind of eggs you eat.

Eggs and chicken meat can also provide you with an additional income stream. A lot of people prefer organic produce, and you can easily find a market for any extra eggs that your chickens produce.

This means that other than feeding you and your family, chickens can also serve as a source of income, and ultimately you may find that the chickens end up paying for their feeds and maintenance cost.

If healthy nutrition sounds good to you, but you are not sure about the kind of costs that will come with raising backyard chickens, that's another area where chickens have an advantage over other types of pests. Chickens are relatively inexpensive to acquire and maintain. For most breeds of chicken, the buying costs per bird will generally range from as low as $3 to $30. This cost will depend on factors such as age and breed, but all things considered, for a pet that is going to provide you with eggs and meat, the startup cost for raising backyard chickens is pretty minimal.

As for chicken feed, this is another area that will not cost you too much. People with modest flocks of six birds or less find that chicken feed only sets them back between $20 and $30 a month. Chicken feed tends to be inexpensive and readily available in feedstores, pet stores, and even grocery stores. There will be differences in the cost of feed based on the brand you choose and the type of feed you go for, but on average, most people find that chicken feed is affordable.

Another plus when it comes to feeding chickens is that they are omnivores and therefore tend to eat most types of foods. Besides the chicken feed, you will find that chickens will happily eat any scraps off your table. This means that any scraps or leftovers from your dinner table need not go to waste.

Chickens are also great foragers. When left to roam in open areas such as yards or gardens, they will dig up bugs and other types of edibles that they find on the ground. Chickens are not fussy eaters, and this means that you do not have to worry about keeping them well-fed and happy. Of course, it is important to ensure they eat healthy stuff since you want them to produce high-quality eggs.

For people with gardens, chickens are not just inexpensive, low maintenance pets, but they also make some of the best natural manure. If you want to fertilize your garden using organic manure, chickens will provide you with one of the best natural fertilizers.

Poultry manure contains plenty of nitrogen, phosphorous, and potassium, as well as other essential nutrients that improve the quality of your soil and give you healthier plants.

Since organic manure is safer for the environment, your crops, and your health, you can simply clean out the chicken manure from the coop and add it to your compost heap. Additionally, if your chickens are free-range, they will effectively fertilize the soil in your garden or yard as they roam about. Since chickens also tend up to dig and scratch the ground for bugs and eat unwanted weeds, they make great garden tillers, especially when you are getting ready to plant or have just cleared your garden.

When you want to get your garden ready for the next crop, your chickens can help with the clean-up and fertilizing of your garden. By using the organic manure from chickens instead of artificial fertilizers and other chemicals, you will find it much easier to grow organic crops in your garden. The kind of products or chemicals that you use in your garden will end up in your plants, and ultimately in your food, so organic manure gives you the option to grow food that is organic and free of harmful chemicals. If you are looking to farm more organically and reduce the use of artificial chemicals and fertilizers in your garden, you will find that chicken manure is an excellent inexpensive alternative.

Chickens may just be one of the most useful pets that you can keep in your backyard. These handy birds are good for more than just eggs and will prove a worthwhile addition to any backyard. If you have thought about raising your own chickens for a while, with a little bit of effort on your part, you can enjoy a host of benefits from your flock. While you will need to put in some effort to raise chickens in your backyard, you will find that the rewards will far outweigh any cons.

Chapter 2: Things to Consider Before Getting Chickens

There are plenty of reasons that make raising your own chickens a pleasurable and rewarding venture. From having a ready supply of fresh eggs to the pleasure of watching your chickens thrive, raising backyard chickens is appealing on many different levels. However, as much as keeping chickens is rewarding, it is still a responsibility that requires time and effort.

This means that before you join the bandwagon of people raising chickens in the backyard, you must be sure you are up to the task. There are important things that you need to keep in mind before you embark on raising chickens. Let's take a look at some of the factors you need to consider.

I. Are chickens allowed where you live?

II. Do you have enough space in your backyard to raise chickens?

III. Do you have the time to raise chickens?

IV. For what purpose are you raising chickens?

V. Are you ready for the costs involved?

VI. Do you have other pets, and if so, can they co-exist with chickens?

Are chickens allowed where you live?

If you are a beginner to raising chickens, before you consider any other factors you first need to ensure that you are allowed to keep chickens in your area. This means you need to find out what the local laws are. The last thing you want is to get on the wrong side of local ordinances and laws.

To find out whether you can legally keep chickens where you live, you can check with your local zoning office or county office. Most towns and cities will have their own regulations on keeping livestock and poultry. Some counties also provide online resources that offer guidance to people looking to keep poultry or other livestock in their backyards.

You may find that you are required to acquire a permit for your chickens. This is more or less similar to the kind of permit you get for dogs or cats.

You may also find that, even if the laws in your area permit the keeping of chicken, there is a limit to the number of chickens you are allowed to keep. This limit will depend on factors such as the size of your land and property lines. However, each county has its own set limit on the number of chickens you can keep. Once you have this information, you will be able to comply fully with the set regulations and avoid any complications down the road.

In some places, the local ordinances are flexible, and you may be able to get a permit for keeping extra birds above the stipulated limit. Another regulation that you need to familiarize yourself with is whether or not you are allowed to keep roosters. Roosters tend to pose noise concerns, and the keeping of roosters is not allowed in some towns and cities. Some areas will allow you to keep a rooster but only up to the age of four months.

You will also need to understand whether your local ordinances permit you to have chickens that roam free in your backyard. Depending on where you live, you may find that there are enclosure restrictions that require you to keep your chickens in an enclosure or within a contained environment. These will be an important regulation to seek clarity on, especially if your aim is keeping chickens as a free-range flock.

In some areas, you may need to get approval on your coop plans and building materials before you can set up a coop in your backyard. Before you start building your coop, check your local laws to see what the regulations are. In some cases, you will find that there are distance regulations imposed to guide you on how far your chicken coop should be from property lines. The distance required from property lines can range from 10 to 90 feet, so make sure you are clear on what your local law stipulates before putting up your coop.

In addition to local laws and ordinances, you will also need to check if there are any regulations on keeping poultry made by your neighborhood residents' association. Since chickens can be a noisy, smelly, hygiene concern for your neighbors, it is always advisable to check for any neighborhood laws that regulate if or how you can keep chickens. You do not want to aggravate your neighbors, so giving them a heads up on your project may help to ensure some goodwill and prevent resistance from the people who live in your area.

Ultimately the laws in your area will determine whether or not you can keep chickens, how many you can keep, and any other regulations. If your local town or city does not allow the keeping of chickens, that does not necessarily spell doom for your dream. People have successfully petitioned their local governments to change their ordinances and laws on the keeping of poultry. You can do this through your local city council. Changing local ordinances or laws may take some time, but if you are patient and consistent, you may end up having the regulations in your area reexamined.

Do you have enough space in your backyard to raise chickens?

While chickens take up relatively little space, you still need to make sure that your backyard has enough room to accommodate a chicken coop and run for your chickens. The general rule of thumb is that you need at least three square feet of space per chicken in your chicken coop. This means that the larger the flock you want to keep, the more space will be required.

Your chicken coop needs to have enough space for the chicken feeders, water containers, and a nesting box as well as a roosting area where the chickens can perch. Chickens spend a lot of time in their coops, so it is important to ensure that it provides a safe and comfortable space for them. When the coop is too small, the smaller chickens may get bullied by the bigger ones. Also, bear in mind that you need to be able to get into the coop to clean it and gather your eggs, so you need to make sure that there is enough room in your chicken coop to stand and work in.

Chickens will also need a run. This is the space on which they can roam and forage. On average, a run of at least fifteen square feet per chicken is adequate, though if you have more space that would be even better. When chickens have ample coop and run space, they are less likely to infect each other with diseases and parasites. Just like you wouldn't want to have any other pet cooped up in a tiny space, ensuring your chickens have enough room is crucial.

If you are looking to raise their chickens as free-range without a run, keeping them in containment, the bigger the space that you have for your chickens, the better. This means that on average, you should work with about 25 square feet per chicken. However, always bear in mind that if your chickens are allowed to roam free in the yard, you need to have safeguards protecting them from predators.

On the whole, chickens will not be too demanding in terms of space. However, before you embark on raising your chickens, set aside the area on which you want to raise your chickens. The size of this area will then guide you as to the number of chickens you can comfortably accommodate in your backyard. Chickens that live in spacious and well- designed areas are ultimately healthier and happier.

Do you have the time to raise chickens?

Having a pet is a responsibility, and chickens are no different. To avoid getting caught up in a hobby for which you are ill-prepared, it is important to understand the kind of responsibilities involved in raising backyard chickens. While keeping chickens has more than its fair share of benefits, there are also chores to contend with, and anyone looking to raise chickens has to be willing to put in the time required.

Most people who keep backyard chickens will tell you that chickens are easy to maintain and do not require round-the-clock attention. However, they do still need to be fed and watered daily, their coops need to be cleaned, and, of course, you will need to collect the eggs. This means that you need to allocate time daily for feeding as well as collecting eggs from your chicken coop.

While thirty minutes a day may not be too taxing for most people, if you travel a lot or are away from your home for extended periods, you will need to have someone taking care of the chickens while you are away. Naturally, the bigger your flock is, the more time you will need to care for your chickens adequately. For beginners, it is always advisable to start with a modest flock. Once you get the hang of the maintenance involved and the ins and outs of caring for chickens, you can then gradually increase the size of your flock.

Chickens tend to poop a lot, and dealing with manure is par for the course for people who raise backyard chickens. This organic manure can get quite smelly if allowed to accumulate, so you will need to find time to clean out your chicken coop regularly. Aim to clean out your coop weekly to prevent the build-up of manure in the coop. Since chicken poop can harbor bacteria such as salmonella, you will need to have protective gear to use when cleaning out the chicken coop. If you have a garden, this organic manure can be used as a fertilizer, so it will also serve a purpose in your garden.

You will also need to clean the waterers and chicken feeders weekly to ensure that your chickens have access to clean, uncontaminated water and food. Thorough cleaning and deep sanitization can be done twice yearly. While several chores will be involved in caring for your backyard chickens, some of these chores only need to be done weekly so they can be easily managed. However, some people do find some of these chores unpleasant, and so before you decided to keep backyard chickens, you need to be sure that you are up to the task.

The health and well-being of your chickens will depend on how well they are cared for. This means that while chickens will provide you with eggs and many other benefits, you will also need to give back in terms of time and effort. Most people get into hobbies without realizing how much time and work will be required, and end up regretting their project. Avoid this pitfall by carefully considering how much time you are willing to spend on raising chickens.

For what purpose are you raising chickens?

People raise chickens for different reasons. Some do it for the eggs, others for the meat, and some do it for pleasure. Whatever it is that you want to raise chickens for will be an important factor when choosing the type and breed of chickens to keep, the size of your flock, and how you choose to raise your chickens.

A chicken, contrary to popular belief, is not just a chicken. There are varied breeds of chickens that each bear unique characteristics. This means that some breeds are better suited for some purposes than others. Chicken breeds vary greatly in terms of temperament, noise levels, egg production capacity, and many other factors.

Some chickens adapt better to confinement than others, so these types of breeds work well for people who are not going to free-range their chickens. Things like noise level and temperament are also important factors to have in mind when determining the best breeds to keep in your backyard.

Chicken breeds that are ideal for people who are mainly keeping chickens for eggs include breeds like Barred Plymouth Rocks and Rhode Island Reds. These two breeds do well as backyard chickens. They are prolific egg-layers and will provide you with a steady stream of eggs. These breeds adapt well to confinement and are generally not noisy, meaning that they will be less of a noise nuisance to you and your neighbors.

Rhode Island Reds also tend to be docile and friendly, so this is a breed that even children can be around and help take care of. In essence, these two breeds check most of the boxes for what you need in a backyard chicken. Another breed that also does well as backyard chickens is the Jersey Giant. It also has a calm temperament. However, Jersey Giants tend to be large and may, therefore, require more space than other egg-laying breeds.

If you want to raise backyard chickens as pets or for pleasure, you will be better off choosing calm and docile breeds such as the Rhode Island Red. Breeds like Araucana are hardier than other backyard chicken breeds but tend to be temperamental and may not make the best pets, especially if you have children. So, before you buy your first flock, always consider the purpose for which you want the chickens. This will help you select the best breed for your needs and avoid disillusionment down the road.

Are you ready for the costs involved?

Raising chickens is a pretty inexpensive venture. However, there are still costs involved, and you need to be ready to inject some cash into your hobby. The initial costs will involve expenses such as paying for permits, buying your chickens, and of course the cost of building a coop and run for your chickens. This means that the higher costs will be at the start of your venture. Once you have everything in place, maintenance costs tend to be largely related to buying feed and getting veterinary care for your chickens if and when the need arises.

At the onset of your project, you will need first to ensure that you have a proper enclosure and housing for your chickens. When it comes to coops, you can buy a ready-made coop or build one yourself. While ready-made coops can save you the time and effort it takes to build one, you will end up spending more money than if you were to build the coop yourself. Online stores such as Amazon have a wide range of chicken coops available that range in price from budget-friendly to costly. This means that you can shop for one that meets your needs but is still within your budget.

When it comes to chicken coops, building your own is usually a popular choice for most beginners. If you are handy with outdoor projects, you can save a pretty penny by choosing to build your own chicken coop. All you need is the building materials and a building plan for your coop, and you are good to go. Some people enjoy putting up their own coops because they can make it exactly how they want it to be in order to best suit their needs.

Another advantage in terms of costs is that when you choose to build a coop yourself, you can easily use recycled material to make the chicken coop. This means that you can make use of whatever appropriate materials you have at hand without necessarily having to buy new building materials. Again, this is a plus if you want a cost-efficient way to start keeping your backyard chickens. Some people find it easier to convert an unused shed into a chicken coop. If you have an outdoor shed that is not used, you may consider converting it into a chicken coop.

The other cost that you will have to meet at the beginning of your project is buying the chickens themselves. Chickens are inexpensive pets, and prices start from as little as $2 depending on the age and breed of chicken you need to go for. If you want to save on the initial buying costs, chicks are generally cheaper than full-grown chickens so you can choose to buy chicks and raise them yourself until they reach egg-laying age.

Once you have your chickens and their coop in place, you will, of course, need to feed them. This means that you will have recurrent costs in terms of buying feed. For egg-laying chickens, the average feed consumption weekly tends to be about 1.5 pounds. This means that with a modest flock, a bag of feed will last you quite some time, and the costs of feeding your chicken will not be high. Chickens are also omnivores, and they tend to eat most things. This means that any leftover food does not need to go to waste since you can feed it to your chickens.

You may also incur additional expenses in terms of veterinary care in case of diseases. Also, plan for costs such as buying feeders and waterers and other miscellaneous items for your coop(s). On the whole, since the chickens you will be keeping will also be providing you with eggs and, for some people, meat as well, the costs and benefits usually tend to balance in favor of the benefits. If you have decided to raise chickens in your backyard, you will need capital investment, but it will not be too high, especially if you just want to keep a small flock of chickens.

Do you have other pets, and can they co-exist with chickens?

Before you bring chickens home, you need to be sure that you have a safe environment for them. If you have other pets, will they be able to share the yard with your chickens? Pets such as cats and dogs may not always be willing to have other animals in their space. They may, therefore, pose a danger to chickens. This is a consideration that you should have in mind, especially if you plan to let your chickens roam free in the yard or live as free-range.

If you have prepared an enclosure for your chickens, make sure it is predator-proof, and yes, this includes ensuring that your other pets will not be able to get at the chickens or harm them in any way. Chickens are susceptible to a lot of predators, and ensuring that they are kept safe will be one of your primary responsibilities. Remember, even friendly dogs or cats can harm chickens, especially if they are still at the chick stage, so always keep them away.

Ultimately, raising chickens is a rewarding hobby, but it is still a responsibility that needs to be taken seriously. Chickens need care and attention to thrive and stay healthy, so before you consider raising backyard chickens, prepare yourself for the responsibility that will come with it.

Chapter 3: Finding the Right Breed for You

The most important decision you will make when you start raising backyard chickens is which breed to keep. For people who have never kept chickens, it may seem like all chickens are pretty much similar. The truth is, however, that there are significant differences between various chicken breeds. This means that for beginners, it is important to understand the key differences between the various chicken breeds and what would make the best match for your needs.

When choosing the best breed, your decision will primarily be based on what you want to keep the chickens for. However, the purpose is only part of what you need to consider. Here are the key factors that should guide your decision on which chicken breed is best for your backyard.

1. Your main purpose for keeping chickens.

2. Your particular climate.

3. The space you have available

<u>Pick a breed based on your main purpose for keeping chickens</u>.

Are you in it for the eggs? While all chicken breeds lay eggs, their production rates and egg size vary from breed to breed. Some breeds are more prolific egg layers, while others are medium egg producers.

If your main purpose in raising chickens is to have a steady supply of eggs for your family and perhaps even a surplus for sale, then naturally, you want to go for breeds that produce the most eggs all year round.

The Best Breeds for Egg Production

- **Rhode Island Red**

This breed is one of the most popular egg-laying breeds in the US and for a good reason. Rhode Island Red chickens can lay approximately 300 eggs annually. This breed lays medium-sized brown eggs. If you are looking for a champion egg layer, Rhode Islands are a safe bet.

In addition to being prolific layers, this breed is pretty low maintenance, which makes it a favorite for people who want to raise backyard chickens. This breed is sturdy, and with good feeding and a comfortable coop, this type of chicken thrives with little attention required on your part. They also tend to have a mild temperament, so they make great pets.

- **Plymouth Rock**

Plymouth Rock is another prolific egg-laying breed that will do well as a backyard chicken. On average, this breed will lay approximately 300 eggs a year. Another reason to choose this breed is that they adapt well to confinement so they can thrive in small spaces.

This breed is docile and makes great pets since they are not aggressive or territorial. They also have striking black and white plumage that makes for a beautiful flock.

- **Australorp**

Australorps are great egg layers and can average up to 300 eggs per year. This large breed requires plenty of space due to its large size. This means they will make a great choice if you have a big backyard with lots of space for your chickens.

- **Black Sex Link**

If you want a beautiful bird that still provides you with plenty of eggs all year round, Black Sex Link may be the right breed for you. This breed produces light brown eggs and can average up to 300 eggs per annum.

These prolific layers are a cross between Rhode Island Red and Barred Rock hen. This breed is popular not just for its prowess in egg-laying but also because it is a hardy breed that will not require a lot of maintenance.

- **ISA Brown**

This breed is also a prolific layer that suits people whose primary goal in raising chickens is egg production. ISA Brown chickens can lay up to 300 eggs per year, putting them in league with the best egg-laying breeds. This breed does well in confinement and will adapt well to backyard living. Since they are so docile, this breed also makes great family pets.

Breeds That Produce Blue Eggs

If you fancy an exotic breed that will give you something other than the common brown or white eggs, some chicken breeds lay blue eggs. These breeds include:

- **Araucanas**

This breed tends to be rare, but makes a great backyard chicken and will provide you with blue eggs. This chicken breed is easily recognizable as it lacks a tail head - a feature common in other chicken breeds.

- **Cream Legbars**

Just like Araucanas, Cream Legbars lay blue legs. However, their eggs tend to come in different shades of blue and not just a single uniform blue. This breed is, however, best if you want to raise free-range chickens. This is because it does not adapt well to confinement and will not thrive in containment or small spaces.

- **Ameraucanas**

This is a distinctive breed that stands out with its characteristic beard. These chickens also lay blue eggs and are a great choice for backyard chicken breeders with a taste for the exotic.

Are you looking for a breed that is ideal for meat production?

If you are raising chickens specifically for meat, then you will find that some breeds are better suited for this purpose. Generally, good meat producers tend to be large breeds that typically grow at a much faster rate than egg-laying breeds. This, of course, does not mean that you will not get eggs from the meat-producing breeds. It only means that they are not as prolific in egg production as the egg-laying breeds.

The Best Meat Producing Chicken Breeds

- **Jersey Giant**

True to its name, this breed grows to an impressive size within 20 weeks. It is a favorite breed among people who keep backyard chickens as a source of meat. They will require ample feed to reach their maximum weight.

- **Freedom Rangers**

This is a common meat-producing breed that grows pretty fast. If you do not have the patience to raise the slow-to-mature Jersey Giant breed, you can choose to keep Freedom Rangers. This breed grows to maturity in about 11 weeks. This makes it a popular choice amongst meat breeders. Freedom Rangers are also reputed to have great tasting meat. However, they do require plenty of space to forage and roam in, so go for this breed if you have adequate space in your backyard.

- **Cornish Cross**

Size is an important factor in selecting the best chicken breeds for meat production. This is one of the reasons the large-sized Cornish Cross is a preferred choice for people raising backyard chickens for meat. This breed grows fast and will reach full maturity in about six weeks. It is renowned for its large thighs and ample breasts, which are good qualities to have in meat-producing breeds.

- **Bresse**

Bresse is not the fastest-growing meat producing chicken breed, but this breed is popular among people who are looking for quality meat. This breed weighs in at approximately 7 lbs. So, it may not be as large as the other meat-producing breeds but is a great choice if you want a breed that is not too large but still adequate for meat production.

Do You Want a Dual-Purpose Breed?

For some people, the ideal chicken breed is one that can be a source of both eggs and good quality meat. If these sounds like just what you need in your backyard chickens, these are the breeds you need to consider.

- **Marans**

Marans are a chicken breed that can serve as both a medium egg-producer and as a source of meat. This breed is available in a variety of colors, including copper blue, black-tailed buff, and golden cuckoo, among others. These chickens are a hardy breed that does not require a lot of care and maintenance. They make great backyard chickens since they adapt well to confinement and are generally mild-tempered. This breed lays dark brown or chocolate-colored eggs.

- **Sussex**

This breed is popular across the globe and is one of the best breeds if you want to raise dual-purpose backyard chickens. They do well as free-range chickens and are good at foraging for food. If you want a chicken that is child friendly, the docile Sussex meets this criterion and makes a great family pet.

- **Wyandotte**

This breed comes in a variety of colors and is often raised as a show bird. However, if you want a breed that will serve as both a source of eggs and meat, the Wyandotte does both well. It also makes a great backyard chicken since it adapts well to confinement and is naturally mild-mannered.

- **Turken**

This breed is also commonly referred to as naked-neck chicken owing to the fact that it does not have any feathers on its neck. This breed tends to be a hardy, low-maintenance chicken that is ideal both as a source of fresh eggs and meat. This breed is generally not fussy and is mild-mannered enough to make a docile family pet.

Which Breeds Make the Best Pets?

Some people keep chickens simply because they want a pet and a pleasurable hobby. If this sounds like you, then you need to know which breeds are ideal as pets. When you are looking for a breed that will make a good family pet, you need to consider the temperament of each specific breed. Some breeds can get quite broody and may attack, especially when they have small chicks.

If you are looking for a docile breed that will be safe even around children, these are the breeds to go for:

- **Plymouth Rocks**

This is one of the most docile chicken breeds and is ideal if you want a family pet; as an added plus this breed is a prolific egg-layer, so you get the best of both worlds if you choose to have this breed in your flock of backyard chickens.

- **Buff Orpington**

Orpingtons make great pets due to their docile and easy-going nature. They are friendly and make calm pets that even children can help to care for. This breed is also dual-purpose, which means it will provide you with a decent supply of eggs and meat if required.

- **Australorp**

What can be better than a friendly pet that also provides a steady supply of eggs for you and your family? If this sounds like just what you need, then the Australorp breed is an ideal choice for your backyard flock. These mild-tempered chickens are friendly and curious, and they do well around people.

- Cochin

Though quite big in terms of size, the Cochins are gentle giants that are usually calm and friendly. This is a bird that enjoys a cuddle. This fluffy bird likes to have some lap time and easily bonds with its owner. Since this breed is also a medium egg-layer, you get a friendly pet as well as a source of fresh eggs if you make this breed part of your backyard chicken flock.

While chickens, on the whole, cannot be considered aggressive pets, there are some breeds that can get quite broody. The Silver Laced Serama is often touted as the most aggressive chicken breed, so you may want to steer clear of this breed if what you are looking for is a friendly family pet.

Bantam Chicken Breeds

Bantam chickens differ from regular chicken in one major aspect: size. A Bantam chicken will typically be roughly a quarter of the size of regular chicken breeds. This small breed of chickens is a good egg-layer and comes with the added advantage of consuming less feed than the other chicken breeds.

If you have a small backyard, Bantam chickens make a great choice since their small size means they can be adequately housed and free-ranged in smaller spaces. Here are some of the reasons you may want to make Bantam Chickens your choice when choosing a breed to raise in your backyard.

I. They are good egg-layers. Each chicken produces about 4-5 eggs a week.

II. They make great pets due to their docile nature and small stature.

III. They require less feed than other breeds making their maintenance cost lower than that of normal-sized chicken breeds.

IV. These little birds are adorably cute and will make for a beautiful flock. Some people even raise them as show chickens.

Consider the best breed for your particular climate.

Different chicken breeds will thrive in different climates, depending on their natural resilience and the adaptabilities they have developed over time. If you are a beginner, it is best to go for breeds that are suited to the climate in your area. This will help to minimize the risk of diseases for your chickens and boost their overall wellbeing.

Best Breeds for Cold Climates

For cold areas, go for breeds that are suited to the cold weather. These breeds will have lots of feathers on their bodies to help keep them warm. Most of them also tend to have feathered legs, which again helps to keep the bird warm. As a natural adaptation to cold climates, breeds that do well in cold conditions will have small combs. This helps them to avoid frostbite.

If you live in a cold climate, these are the breeds you should aim to make part of your backyard flock.

- **Rhode Island Reds**

This prolific egg-laying breed is well suited to cold weather and will do well in cold climates. Their plush feathers effectively keep this breed of chickens well insulated from the elements.

- **Australorps**

Just like Rhode Island Reds, Australorps have heavy plumage, which helps them stay warm even in cold conditions. When you go for birds that are adapted to colder climates, you are sure that they will thrive in your backyard and you will not have to deal with constant ill-health or even birds dying due to adverse weather conditions.

- **Brahma**

This breed has the characteristic feathered feet that make some chicken breeds better suited to cold climates. This big chicken breed is docile and very friendly and therefore makes great family pets. It will also keep you supplied with eggs, although it is better known as a meat-producing breed.

If you are in a cold climate and want a hardy breed that is built to withstand cold conditions, Brahma chickens are a good choice for your flock.

● **Dominique**

This chicken breed is small in stature but is equipped with enough plumage to keep it warm in colder climates. In fact, this breed does not tolerate heat well and is therefore ideal for you if you live in a cold region and need a chicken flock that is suited to that particular climate.

● **Ameraucanas**

Ameraucanas are famous for their blue eggs, but this particular breed also thrives in cold conditions. Although it is not the most prolific egg-layer, you will still get a decent supply of eggs from this breed.

Best Breeds for Hot Climates

If you leave in a hot climate, then similarly, you will need chicken breeds that are suited to that particular climate and can survive the high temperatures. In terms of raising chickens, areas that are classified as hot are those that average 89.6 degrees Fahrenheit or higher. For breeds to do well in such hot weather, they need to have natural adaptations that reduce the effect of the heat on the chicken. These adaptations include lighter plumage, lighter colors that do not absorb as much heat, and smaller bodies.

For hot climates, these are the breeds that will thrive and do well.

● **Plymouth Rock**

We have already covered this particular breed under the best breeds for egg-laying as well as its suitability as a family pet. These attributes make it one of the most popular backyard chicken breeds for hot areas. This hardy breed is adaptable and will do well in both cold and hot conditions, making it one of the most versatile breeds you can have in your flock.

- **Golden Buff**

This breed is hardy and adapts well to hot climates. It also does well in cold climates so you can keep it regardless of what kind of climate you live in.

- **Leghorn**

For hot climates, the Leghorn breed stands out due to its sturdy and resilient nature. These birds are good egg-layers and are recommended for people who want a breed that is good for egg-laying and thrives in hot climates.

- **Fayoumi**

Fayoumis are striking birds that are hardy enough to thrive in extreme heat conditions. They are well adapted to hot climates and would suit you if you want a striking flock of exhibition chickens.

Pick breeds based on the space you have available.

Large breeds will naturally require more space, and therefore you should be aware of the size of the breed you are buying. Often, if you are buying chicks you may not be able to estimate the potential size of the full-grown chicken. While space will only become a pressing issue if you want to keep large flocks, it is still best to ensure that you have adequate space for the particular breed you want to purchase.

When chickens get overcrowded in small spaces, the risk of infectious diseases and parasites spreading amongst them increases significantly. This can end up being costly, and it is therefore best to avoid the situation altogether. Most breeds that are best for meat production tend to be larger than egg-laying breeds, which means that if you want to raise chickens for meat purposes, you will probably need more space.

Large chicken breeds that require plenty of space include breeds such as Jersey Giant, Cochin, Brahma, Cornish, Orpington, Rhode Island Red, and New Hampshire. Though these breeds are large, it is still possible to breed them on a small yard as long as you stick to a modest flock to allow each bird enough living space.

It is also important to note that apart from more space, raising large chickens is pretty much the same as raising smaller and medium-sized breeds. Some people mistakenly think that larger breeds are more aggressive. This is not true in any way as a chicken's temperament is not connected to its size. In fact, most of the more friendly and docile breeds tend to be large breeds like the Jersey Giant, Rhode Island Reds, Cochin, and Plymouth Rocks.

Ultimately, whatever breed of chicken you choose for your backyard flock, you will need to care and nurture for them well to enable them to thrive. Taking good care of your backyard chickens by ensuring that they are well-fed, housed adequately, and have a clean, safe environment will be the main factors in determining whether or not you get the best out of your backyard chicken flock.

Chapter 4: Getting Set Up and Selecting a Coop

Before you bring the chickens home, you will need to set up their living area. This means having a coop to shelter your chickens and a chicken run for them to forage in, as well as all the other materials required for the proper care of chickens. If this is your first time raising chickens, then you probably need to start your setup from scratch. This means identifying where you want your chicken coop to be, how large or small it is going to be, and whether or not you will free-range your chickens.

Chickens, just like any other pet, have basic requirements that need to be met. These requirements are the factors that will guide you on how to prepare your backyard for your chickens before you bring them home. You want to avoid a situation where you bring your flock home only to realize that you are missing something essential.

These are the basic requirements that are necessary for your chickens to be healthy and happy.

- A well-constructed shelter to house the chickens, which is the chicken coop
- Food and water
- Enough space to move around

- A chicken run or forage area for them to dig, scratch, etc.
- A safe nesting place for broody hens

Choosing the Right Location for the Chicken Coop

When it comes to preparing to raise backyard chickens, the first thing you need to figure out is the coop, which is basically where your chickens will be sheltered. The chicken coop poses several questions that you need to answer before you can identify the right coop for your particular needs; things like how big a flock you want, whether to build or buy the coop, the appropriate size, and whether or not you want a stationary coop. However, even before you get to all that, you must first figure out where to position the chicken coop in your yard.

Location is an important consideration when it comes to providing appropriate shelter for your chickens. Where you place your chicken coop entails factors like how much sun and shade your chickens will get, wind exposure, safety, convenience, and a host of other key factors. Besides, chicken coops can get quite smelly, and they also tend to attract insects. This means that if you position the coop too close to your house, you may have to contend with unpleasant odor and bugs.

To ensure you get the location of your coop right, keep the following considerations in mind as you pick the ideal spot for your chicken coop.

1. Distance from the Chicken Coop to Your House

The general rule is to ensure that you do not place the chicken coop right next to your house. Since chicken poop tends to have a strong odor, this smell can easily become a nuisance if the chicken coop is too close to your house.

You will also find that chickens tend to attract insects and bugs, which you, of course, do not want to become a permanent fixture in your home. When choosing the best location for your coop, identify a spot that is not directly next to your house but is still close enough to access conveniently.

Having the coop not too far away from your house means you can easily check on them when needed. Since you also need to feed, water, and collect eggs from the coop, this means you will make trips to the coop daily, so having it close by will make your chores a lot easier.

In case you need to connect electricity to your coop for heat or any other reason, it will be easier if it is not too far away from your house. This also goes for amenities such as water for cleaning and watering your chickens. In a nutshell, find a location for your coop that is not directly adjacent to your house but is also not too far off.

2. Find a Location with a Level Ground

It is important to ensure that the chicken coop is situated on level ground. This will help to ensure that the structure is stable and durable. You can clear a level patch for your chicken coop. Remember that the area will also need to have good drainage since you do not want your coop getting submerged in water during the wetter months.

If you live in an extremely wet area putting in a concrete foundation will make your chicken coop structure more stable and durable. Some coops are constructed with floating floors, which basically means that the floor is suspended above the ground on concrete blocks to create a level surface.

3. Your Coop Should Have Enough Foraging Area Around It

Chickens love digging around for bugs, foraging, and looking for food on the ground. You therefore need to ensure that your coop has sufficient forage area around it. The size of the forage area will, of course, depend on the size of your flock, but having at least 8 square feet per chicken is best. An ideal forage area can be a combination of a grassy patch and dirt.

When you do not allow enough space for chickens to roam, they become more prone to infections and ill health. If you are confining your chickens to a run, make sure it has sufficient space for them to forage based on the size of your flock. For free-range chickens that are not confined to a run, you will still need to make sure your yard has enough forage area for the number of chickens you intend to keep. For free-range chickens, the recommended forage area is approximately 250 square feet per bird.

4. Your Coop Should be in an Area That is Not too Windy

You want your chickens to be nice and warm in their chicken coop, especially during the colder months. This means that when picking the location for your coop, consider whether the area has a windbreak. Positioning your coop in an area with some windbreaks such as trees or a tall structure will ensure that the temperatures in the coop are not adversely affected by windy conditions.

5. Pick an Area That Gets Some Sun but Also has Some Shade

Your coop should be in an area that gets some sun. It should also have some shady areas where your chickens can seek respite from the sun in hotter months; chickens thrive in a setup where they can enjoy both some sun as well as some shaded areas when it gets too hot. If you can find a sunny spot with a few trees that can offer some shading for the coop and forage area, that will be an ideal location.

How to Choose the Right Chicken Coop

Once you have figured out the perfect location for your chicken coop, the next step, of course, is to identify the best chicken coop for your needs. There are plenty of options when it comes to chicken coops. The variations in size, shape, and design mean that there is a coop available to suit different requirements. However, you do need to know what makes a good chicken coop before you settle on a particular design.

These are the main considerations to keep in mind when choosing a chicken coop.

I. Size

II. Internal structures; roosting bars, and nesting boxes

III. Ventilation

IV. Safety

Size

The first consideration when choosing a chicken coop is to ensure that the coop is the right size for your flock. Depending on how many birds you want to keep, and the size of the breed you have chosen, the amount of space you need in the chicken coop will vary.

You must ensure that your coop is the right size for the flock you intend to keep. While you can allow more square footage than the recommended guidelines, do not go overboard. A coop that is too large may be colder and require extra heating to keep your chickens warm.

For large breeds, such as Jersey Giants, Plymouth Rock, or Rhode Island Reds, you will need a minimum of 4 square feet per bird in the coop. This is the minimum space per bird, and you can always have a higher allowance of space for your chickens.

For medium-sized breeds such as the Leghorn, you should have a space allowance of at least 3 square feet per bird. Again, this is the minimum space, so you can always have a higher space allowance for your chickens.

Smaller breeds such as Bantams do not require a lot of space. An average of 2 square feet per chicken should be enough if your flock is made up of small breed chickens.

When your coop is too small for your flock, it can cause the following problems:

- Poor chicken health due to high ammonia levels in the coop from accumulated chicken manure.
- Poor egg production due to cramped conditions in the coop.
- Bullying and aggressive behavior among the flock as they each fight for space.

Internal Coop Structures: Roosting Bars and Nesting Boxes

Once you have calculated the square footage of the coop that will be adequate to house your flock of chickens, you then need to consider the appropriate size of the structures inside the chicken coop.

One of the most important structures inside your chicken coop will be the roosting bars. Chickens do not sleep on the ground. They need roosting bars raised from the surface of the coop. The roosting bars need to be higher than the nesting boxes in the coop.

The roosting bar needs to provide adequate room for each bird inside the coop to avoid overcrowding. Your roosting bar should provide approximately 8 inches of roosting space per chicken. During colder months, chickens tend to draw closer to each other, so do not make the roosting bars too big.

Nesting boxes provide a private space for your chickens to lay eggs and for broody hens. While you do not need to have too many of them, ensure that you have at least one for every three chickens in your flock. This means that for a flock of 12 chickens, four nesting boxes will be sufficient. Having enough nesting boxes ensures that your chickens have a safe place to lay their eggs.

The Chicken Run

Apart from the inside of the coop, you will also need to ensure that the outdoor space for your chickens, or the "chicken run", is the appropriate size. Remember that when your chickens are not in the coop, they will be foraging outdoors in the run, so it is an important extension of the chicken coop. The recommended square footage for the chicken run is at least 8 square feet per bird.

If you plan to free-range your chickens, you may not need an outdoor run. However, even for free-range chickens, having a confinement area may come in handy when you need to pen the chickens up for a while for one reason or another.

Ventilation

The chicken coop needs to be properly ventilated, and it should allow sufficient air circulation. This means that your coop should have enough air vents to let air in and out of the coop. When ventilation is poor, the build-up of ammonia in the coop from the chicken poop can cause ill health to your chickens. Make sure the air vents are well secured with chicken wire to prevent predators and rodents from getting into the chicken coop.

Safety

Unless you want to wake up one day and find that a wily fox or raccoon has fed on your chickens, you will need to make safety a priority when choosing a chicken coop. Chickens have plenty of predators, ranging from the average domestic dog or cat to possums, foxes, and raccoons. This means that your chicken coops should offer adequate protection for your chickens.

A secure chicken coop should not have gaps and spaces that can allow rats or snakes to enter the coop. Even the ventilation spaces should be well covered with chicken wire to keep rodents and predators out. Also, ensure that the doors are well secured with childproof locks that stay locked. Keep your chickens locked into the coops every night.

It is always advisable to keep your chicken feeds in a separate area and not in the chicken coop. This is because some predators will be attracted to the coop because of the chicken feed. The same should also go for chicken eggs. Do not get into a habit of leaving them uncollected in the coop for days, as they will attract predators to the coop. Some predators are more interested in the chicken feed and the chicken eggs than the chickens themselves, so if you can avoid putting feed in the coop, you will deter such predators.

Your chicken coop should also have a secure roof. This will serve to keep predators out and also ensure that your chicken coop is kept moisture-proof, especially during wet weather. Chickens do not like to be rained on, so providing them with a well-covered coop is important.

Safety precautions will also be needed for the chicken run since your chickens will also be spending time outdoors. Your run should be enclosed using fencing materials that have very small openings such as chicken wire, welded wire mesh, or electric netting. This will ensure that even when your chickens are outdoors, they are well protected. In areas with flying predators such as hawks and owls, overhead fencing can be used to make your chicken run more secure.

If you are looking to raise your chickens as free-range, you will need to make sure that your yard is well secured so that your chickens can forage safely. This means making sure that your yard fencing is intact and high enough to keep out predators. You can also bury welded-wire mesh or any other small mesh fencing material to deter digging predators that tend to dig holes to access the chickens in the run.

Ensuring your Coop is Chicken-Ready

Feeders and Waterers

Once you have found the right location for your coop and found the best coop for your needs, the next step is to make sure that your coop is equipped to shelter the chickens. Naturally, you will need to feed your chickens, so you will need to get chicken feeders and waterers for the coop.

Chicken feeders come in a variety of shapes and sizes. The best feeder for your chickens will depend on the size of your flock. There are also smaller feeders that are designed to be used specifically for chicks, so make sure you buy the appropriate feeder.

The most common type of chicken feeder is the dispenser feeder. This feed dispenser gradually releases feed as it is consumed. These types of feeders can be strung up to keep dirt and debris out of the feed and also to discourage bugs and rodents.

Some people choose to go for automatic feeders that only need to be refilled occasionally. This can be a good choice if you want a feeder that does not need to be refilled every other day. However, you do not necessarily need an expensive feeder to get the job done. Even a homemade chicken feeder from recycled materials such as plastic containers will serve the same purpose as a store-bought feeder. You can easily make a homemade chicken feeder using a bucket or any other type of plastic container.

It is always advisable to keep your feeders inside the coop to protect the feed from the elements. However, you can still choose to have a feeder in the chicken run provided you position it where it is safe from rain, dirt, and debris. To make sure all your chickens have access to a feeder, ensure that you have enough feeders to accommodate the entire flock. If your feeders are too few, the smaller chickens may get bullied during feeding time. Have at least one feeder for every ten birds.

Apart from a chicken feeder, you will also need a waterer or drinker for your chickens. Chickens need to stay hydrated, and they must have access to clean drinking water. Waterers, just like feeders, come in a variety of shapes, sizes, and designs. Consider the size of your flock when selecting a waterer. Some waterers need to be wall-mounted, so only go for this type if the coop you have can accommodate a wall-mounted waterer.

The most common types of waterers are gravity fed waterers, automatic waterers, and container waterers. Gravity-fed waterers are popular because they are easy to use and, therefore, the most convenient. Automatic waterers are ideal if you are short on space since they come with cup or nipple attachments for the chickens to drink from. You may need to train your chickens on how to drink from the automatic waterer, but most tend to get the hang of it pretty fast.

Most waterers are made out of either steel or plastic. Both of these materials are durable. However, steel waterers can heat up considerably in hot weather and may not be ideal if you live in a hot climate. Plastic waterers are generally cheaper than the steel ones, so they are a good option if you are on a budget.

Keeping your waterer inside the coop can lead to wet bedding. To avoid this situation, most people choose to put their waterers outside the coop in the chicken run. You can also reduce the risk of spillage and leaks by not overfilling your waterers. Just like with a chicken feeder, ensure that you have enough waterers for the size of your flock.

Chicken Coop Beddings

Chickens poop a lot, and having bedding in your coop will help to keep the coop clean and odor-free. The bedding that you will use on the floor of your chicken coop will serve as litter to aid in controlling odors and moisture in the coop. Bedding will also provide insulation for the coop, so it is good practice to put bedding in your coop before you bring your chickens home.

The best type of bedding for a chicken coop is an absorbent material that will aid in keeping the coop floor dry. A wet coop floor can lead to diseases and may cause lesions on your chickens' feet. So, when picking the best bedding for your coop bear in mind that you need to have a material that will absorb and release moisture quickly. Common bedding materials include wood shavings, straw, hay, and grass clippings.

Wood shavings make one of the best options for chicken coop beddings. They are absorbent but do not retain moisture for long, so your chicken coop floor stays dry. You can put newspapers under the wood shavings for easier cleanup but do not use newspapers alone as bedding in your chicken coop.

For the nesting boxes, you can use straw and hay to cushion your chickens when they are laying eggs. This will also ensure your eggs do not break. Alternatively, you can also use the same wood shavings you have used on the rest of the coop floor in your nesting boxes.

Ultimately the whole setup of the chicken coop and chicken run should be centered on providing your chickens with a clean, safe, and comfortable living environment.

Chapter 5: Building a Chicken Coop

Once you have set up your yard and decided on the kind of chicken coop that you will need for your backyard chickens, you have three options; buy a pre-made chicken coop, repurpose an existing structure, or build one yourself. Depending on your budget, the kind of design you want, and how handy you are with DIY projects, the choice will be easy to make.

Some people, especially those with big pasture areas, opt to have portable chicken coops. This type of coop can be moved from one section of your land to another, in effect allowing the chickens to forage on different sections of pasture. These types of coops work well if you have plenty of space and pastureland. However, for small backyards, especially in towns and cities, a stationary coop may work better since it does not require repeated moving.

If you do not have the time or the know-how to build your own chicken coop, you can always go for a pre-made chicken coop. These are widely available in online stores such as Amazon, as well as in pet shops and even grocery stores like Walmart. Pre-made coops come in various sizes, designs, and price points so you can go for one that falls within your budget while meeting all the requirements required for the number and type of chickens you plan to keep.

Repurposing an existing structure is also a simple way to create a chicken coop. If you have a shed that is no longer in use, you can repurpose it to serve as your chicken shed. All you will need to do is outfit it for chickens by adding nesting boxes, roosting bars, and some bedding on the floor. Ultimately, this beats building a chicken coop from scratch and is, of course, much cheaper than buying a pre-made coop.

The third option is building your own chicken coop. This gives you the freedom to make a customized chicken coop that suits your backyard and your needs perfectly. You have a choice of getting designs done by a professional or doing them yourself. There are also plenty of online resources that offer free chicken coop plans. Before you settle on a design, always check if there are any town ordinances or neighborhood regulations that need to be met.

If you choose to build your own chicken coop, there are many ways to do it depending on the kind of chicken coop you have in mind. You can choose to have it done by a professional if you are short on time or you do not have access to the materials required. Alternatively, if you want a simple and quick way to do it, you can follow our simple step by step guide for building a 24-square-foot coop that can hold between 6 and 8 chickens.

Materials and Tools Required

- Wood
- Hammer
- Saw
- Measuring Tape
- Pencil
- Screwdriver
- Electric Stand Saw
- Extension cords

- Spirit Level
- Sandpaper
- Paint Brush

While coops can be built using a variety of different materials, wood is the easiest material to use. Wood is also good for insulation, and it makes solid and stable structures that are durable.

Step by Step Coop Building Process

Once you have your supplies ready, you can start building.

1. Start by Building the Floor of Your Coop

- Start with a piece of plywood cut to 4 feet wide and 6 feet long.

- The plywood should be at least half an inch in thickness. This will ensure that your floor is sturdy.

- To make your floor frame, you will need battens. Ideally, these should be 2 x 4s. Screw the 2 x 4s around the borders of the plywood. Also, screw another 2 x 4 across the middle of your plywood floor.

2. Build the Solid Wall Next

- The solid wall of your coop is the one that will not have a window. Take a ½" (or thicker) piece of plywood 6 feet long. You will need 2 x 2s for this wall. Secure the 2 x 2s to the bottom of the vertical edges of your plywood. The 2 x 2s should stop 4 inches above the bottom of the plywood.

- Once you have screwed on the 2 x 2 s, you can now secure the wall to the floor you built in step 1. Take your solid wall and place it on the floor in such a way that the 4 inches that you left cover the 2 x 4s on the underside of the floor. Once you have positioned the wall, screw it in place. Your screws should be 1½" in order to secure the wall firmly to the floor.

3. The Next Step is the Front Panel

- Attach a four-foot length of ½" (or thicker) plywood to the floor and the solid wall you have already built. First, screw the piece of plywood to the two 2x4s at the bottom of your coop; then secure the plywood to the solid wall by screwing it on to the two 2x2s on the solid wall.

- Once the plywood is secured to the coop, it is time to cut the door.

- The door opening should be 2-to-3 feet in width. The height can vary, provided you leave a minimum of about 6 inches between the edge of your door and the bottom part of your plywood panel. The same 6-inch allowance should be left between the edge of the door and the top of the plywood panel.

- Once you have marked out the measurements for the door opening, cut it out with your saw. Make the cut as smooth as possible.

- You will want to reinforce the top of the door opening by using a 20-inch piece of wood. Attach this piece of wood to the top using screws and some construction glue.

4. Let's Build the Back Wall

- Just like the front panel, for the back wall, you will also need a piece of plywood that is 4 feet in length and at least ½ an inch in thickness.

- Secure the piece of plywood to the back of your coop by screwing it to the 2x4s on the underside and then screwing it to the 2 x 2s on the solid wall of the coop.

- Once the back wall is secured to the coop, you can now measure out the door opening for this wall. Using the same measurements that you used for the opening on the front panel, cut out the opening just as you did for the front panel.

- Finally, reinforce the top of your door opening with a piece of wood just as you did to the opening on the front panel.

5. Build the Last Wall

- Cut two pieces of ½" (or thicker) plywood to a length of 2 feet. Next, cut a piece of plywood 5 feet long. The width of this piece should be half the height of your coop.

- Once you have these three pieces of plywood, you can start securing them to the coop to build the last wall. Start with the 2-foot-long pieces of plywood. Secure a 2 x 2 to one of the vertical edges of the plywood. The two 2x2s should stop at least 4 inches above the bottom of the plywood.

- Take the second 2-foot length of plywood and also attach a 2x2 to one of the vertical edges of the plywood. The 2x2s should leave a 4-inch allowance to the bottom of the plywood.

- Now take one of these plywood panels and attach it to the front of the coop. Once that is done, take the second panel and secure it to the back of the coop.

- Now take the 5-foot-long piece of plywood and secure it between the two panels you have just attached.

- The edge of the 5-foot plywood should line up with the tops of the other two panels.

- The next step is to take a piece of wood that is the same vertical length as the middle piece. Screw this piece of wood to the joint where the middle panel connects to the side panel. Do the same for the second joint where the middle panel connects to the other side panel. This way, you have two pieces of wood, reinforcing the two joints between the middle panel and the other two panels.

6. Constructing the Roof

- For the roof, you will start with the gables. These are the triangular structures that will be placed on top of the walls of the coop to support the roof.

- To fit properly on the walls, you need to make your gables 4 feet long. Make sure that the pitch you create for both gables is the same so that your roof seats evenly on the coop.

- The gables will go on the top of the front and back walls. Take the first gable and secure it to the inside of the front wall. Use screws and some construction glue to attach it securely.

- The second gable should be attached to the inside of the back wall. Make sure that the attachment is secure.

- Once the gables are attached, you need to build support for the middle of the roof, which is the truss.

- The angle of your truss should be the same as the one on your gables.

- To make sure you get the right angle, take two pieces of 2 x 2s and clamp them to the edges of one of the gables. The 2 x 2s should be longer than the edge of the gable by about 3 inches.

- You will need a crosstie to reinforce your truss. This crosstie needs to be of similar length to the gables.

- Attach this crosstie to the 2 x 2s with screws. Once it is attached, you can remove the truss from the gable by removing the clamp you had used to attach it to the gable.

- Now place the truss in the middle of the coop.

- Make marks where the 2 x 2s of the truss intersect with the side walls. These marks represent where you will make the notches on your truss.

- Once the notches are made, you can now place the truss on top of the side walls. It should be at the center of the two side walls.

- Now that the roof supports are in place, you need to make the actual roof.

- Using two pieces of plywood, make a roof by joining one 40-inch piece of plywood with an 84-inch piece of plywood. The joints should be along the longer 84-inch sides. You can easily join these two pieces using hinges.

- The roof is now ready to go on top of the coop. It will have overhangs on both sides of the coop.

- You will need to attach two pieces of 2 x 2s to the bottom edge of the front and back overhangs of the roof.

- Once the trim is in place, the final part of constructing the roof is to secure it to the gables on each side and the truss in the middle.

- You can then make your roof moisture-proof by covering it with tar paper or galvanized roofing.

7. Building the Coop Doors

● Now that the walls are finished, it is time to build the doors.

● Take a medium density fiberboard and cut it to the same length as your door opening and half the width of the door opening.

● Construct your door frame using 2x2 pieces of wood. Fasten these pieces on the four sides of your door opening.

● Once the frame is in place, you can now screw in the hinges. Use two hinges for each door.

● Once the hinges are attached, you can now fix the doors to the frame.

● You will then construct the back doors using this same process.

● Once all the doors have been attached to the coop, you now need to put in locks so that the coop is lockable. Get secure locks for your doors that will effectively keep predators out of your coop.

8. Make Legs for Your Coop if You Want it Raised

If you want your coop to be raised, you will need to attach four pieces of 2 x 4s to the underside of the chicken coop. You can secure these legs to the 2 x 4s on the bottom of your coop.

If your coop is raised, you will need a ladder to make it accessible to the chickens. For your ladder, attach 2 x 2s to 2x4s to the required length. Take your ladder and then secure it in place with some hinges.

9. Roosts and Nest Boxes

The interior of your coop will need two essential structures. These are roosting bars and nest boxes. Roosting bars are basically raised bars where your chickens will perch and sleep at night. Chickens do not sleep on the ground, so they need roosting bars that are elevated off the floor of the coop.

For your roosting bars, allow at least eight inches of space per chicken. You can put in multiple roosts depending on the size of your flock. For your roosts, you can use sturdy pieces of wood secured to the coop wall at an angle or even a short ladder propped up at an angle. The roosting bars need to be at least two feet off the coop floor.

The other essential structures to have in your chicken coop are nest boxes. These boxes provide your chickens with a private area to lay eggs. Nest boxes will also be used by the brooding hens when they want to hatch eggs. On average, you will need one nesting box for every four birds.

You can construct one-foot square wooden boxes and use them as nesting boxes in your chicken coop. Alternatively, you can easily repurpose old milk crates and use them as nesting boxes. Whichever type of nesting boxes you choose, simply secure them to your coop walls, and you are all set.

10. Coop Bedding

The last step to prepare your coop and make it ready for your chickens is the bedding. Bedding is an absorbent material that is put on the floor of the chicken coop. The bedding helps to keep the coop floor dry by absorbing moisture from the chicken coop. It also absorbs odor from chicken manure, and this helps to prevent ammonia build up in the coop. Another advantage of having bedding in your coop floor is that it makes clean up easier. It also provides additional insulation for the coop, helping to keep your chickens nice and warm, especially during the colder months.

Wood shavings make great bedding since they absorb and release moisture. Other materials that can be used as bedding are straw, hay, sand, and grass clippings. Be sure to also cushion your nesting boxes with some straw and hay.

Tips for Your Chicken Run

If you do not intend to raise free-range chickens, then you will need to build a chicken run to keep your chickens confined when they are outdoors. Chickens need a foraging area to dig and forage outdoors, so if you do not want them roaming all over your yard or garden, you will need to confine them to a particular section in your yard.

Typically, the chicken run should be adjacent to your coop so that your chickens can get in and out of the coop from the run. Basically, once you have decided the appropriate area that is sufficient for the size of flock you are planning to keep (work with a minimum of 5 square feet per chicken), you can fence this area off to make your chicken run.

To make sure that your run will keep your chickens safe from predators and also keep them from wandering, you should use chicken wire, welded wire mesh, or electric netting as your fencing material. When you use small mesh fencing materials, you effectively keep your chickens safe from smaller predators that can reach through a larger mesh or even jump through it.

If your area has plenty of flying predators such as hawks and owls, you can choose to cover your chicken run with chicken wire or any other small mesh fencing wire. Remember that you want your chickens to be safe in their run. You also do not want to have to keep checking on them constantly to see if they are safe. This means that taking all the safety precautions necessary when setting up your coop and chicken run will save you time in the long run and save you plenty of predator trouble down the road.

Chapter 6: Tips on Buying Chickens

When your coop is all done and you are eager to bring your feathered friends home, then it is time to buy your chickens. Of course, by now you already have an idea of the number of chickens you want, and what kind of breed is best for you. With this in mind, you can now start shopping for the right birds for your backyard.

Where Can You Buy Chickens?

There are plenty of options for people who want to buy poultry. Whether you are looking for day-old chicks, pullets, or mature chickens, there are plenty of hatcheries, feed stores, poultry associations, and breeders where you can buy your flock. It is important to look for reputable hatcheries and breeders so that you can be sure that you are getting birds that are in good health.

If you are not aware of any breeders in your area, check with any farm stores in your area. Usually, they will have information on local breeders or hatcheries. Alternatively, most breeders have some sort of online footprint, so you may get some leads to reputable breeders in your area by checking online sites and social media platforms. If you want a specific breed, you can also check on social media for breeder groups that specialize in the particular breed that you have in mind.

If you choose to go with an online breeder or hatchery, always check consumer reviews and feedback to ascertain that the supplier is credible. Some people prefer to buy their chickens from other farms since they can see the kind of environment the chickens have been bred in. As much as possible, when buying chicks, pick a hatchery or breeder that is closer to you to minimize the amount of time your chicks have to spend in transit.

Farm stores are easily accessible to most people and are a popular place from which to buy chicks. However, when you buy from a farm store, you will not know whether the chicks are male or female, so you may end up with roosters that you did not want. You may also not be able to get information on whether the chicks have been vaccinated or not.

The other alternative is buying from a hatchery. Hatcheries usually have a variety of breeds available and tend to be cheaper than breeders. Hatcheries tend to specialize in utility birds and may not really be a good source if you want heritage chickens. For rarer breeds, breeders are usually a better source. Breeders tend to specialize in particular breeds. They can be a bit pricey compared to hatcheries, but on the upside, you can get even heritage chickens from breeders.

If there are no hatcheries or breeders near you, some hatcheries offer shipping options across the country and will get your birds to you wherever you are. These include:

- **My Pet Chicken**

This hatchery is great for beginners looking to start with a small flock. You can order as few as three chicks. They also carry different breeds, so you have a wide range of options to choose from. As a plus, they sell other accessories and chicken equipment that may come in handy for beginners such as coops, chicken fencing, feeders, and more.

- **Cackle Hatchery**

This hatchery in Missouri has all types of chickens on offer. From layers to meat breeds and dual-purpose birds, you get most types of breeds from this hatchery. Since they allow even small orders, you are not required to buy in bulk, so this is also a good place for the urban dweller who wants to keep a modest flock.

- **Murray McMurray**

This hatchery is located in Iowa and has a wide range of chicken breeds to choose from. They also have various equipment and accessories for chickens, so this can be your one-stop-shop for your chicken needs.

- **Freedom Ranger Hatchery**

Freedom Ranger is ideal for people looking for free-range chickens. This hatchery uses eco-friendly farming methods, and it is well known for its organic free-range poultry.

- **Meyer Hatchery**

This hatchery has over 160 poultry breeds for buyers to choose from. They offer gender-guarantees, so for people who want to buy chicks that are strictly female or male, this is a great hatchery to buy from.

- **Ideal Hatchery**

This Texas Hatchery guarantees 100% live delivery to its clients. They also have plenty of breeds to choose from. However, they do have a minimum order requirement, so they may not be the best option if you want a small number of chickens.

- **Stromberg's Chicks**

This hatchery has locations in five states, including Minnesota, California, Texas, Pennsylvania, and Florida. They have an impressive selection of chicken breeds to choose from, and if you are looking for accessories and equipment as well, they have plenty of those, too.

Information You Need from Your Breeder

Once you have identified the breeder or hatchery you will be buying your chickens from, here are some basic questions you need to ask the breeder.

1. Find Out What Breeds are Available

Most breeders will specialize in a few select breeds, so you need to know what type of breeds they have. You can then decide whether the breeds they have are what you require or move on to a different breeder.

2. Find Out if They Have Sexed Birds

If you are buying chicks, it is not physically possible to tell whether the chick is male or female at that young age. Sexed birds are those that have been checked by the breeder and ascertained to be either male or female. This identification is important, especially if you live in an area where roosters are not allowed.

When you buy sexed birds, you know exactly what you are getting, and there will be no nasty surprises later when one of your chicks turns out to be a noisy rooster.

3. Find Out if the Breeder is Certified by NPIP

Breeders who are certified by the National Poultry Improvement Plan are those that have assented to having their chickens checked for diseases. If the breeder is certified, you will have some assurance that the birds you are getting are in good health.

4. Find Out if the Chickens Have Had Any Vaccinations

When buying chicks, make sure you find out if they have been vaccinated and what type of vaccination(s) were given. This will guide you on whether you need to get any vaccinations done.

5. Find Out Any Specific Care Information for the Particular Breed You are Buying

Breeders can be a great source of information on caring for your birds, especially if the breeder has been doing it for years. Get information on things such as climate needs/ preferences, breed temperament, ideal feed, average egg production, and any other details you are unsure about.

The more you know about the chickens you are buying, the better equipped you will be to take care of them. So, do not hesitate to get as much information as you can from the breeder.

What to Look for When Buying Baby Chicks

The last thing you want is to buy chicks that are unhealthy or poorly developed. One infected bird can easily spread diseases to the rest of your flock. Therefore, before you take any chicks home, make it a point to check for any signs of ill health.

So, how do you know if the chicks you are buying are healthy? Here are some signs to look out for when buying chicks.

I. Eyes should be clear and alert.

II. Check the abdomen for any sign of distension.

III. The chick should be steady on their feet.

IV. The chick should be active and peeping.

V. Check that the top of the beak is aligned with the bottom.

VI. If the chick appears too small or runty compared to others, they may be in poor health.

VII. Healthy chicks are fluffy.

VIII. The vent should be clear of feces or any redness as this could point to diarrhea.

IX. Check to see that the brooder is clean.

What to Look for When Buying Mature Birds

A healthy bird is crucial, especially if you are just starting out in raising backyard chickens. Bringing home birds that do not lay eggs, thrive, or even end up dying can be a disappointing way to start your venture. This means being on the lookout for any signs of ill-health can save you lots of trouble later on.

When buying mature chickens, there are physical signs that can indicate that the chicken is in poor health. These include:

- Any discharge from the eyes or nostrils. A healthy chicken has clear eyes.
- Droopy or swollen eyes. A healthy chicken has clear eyes that are lively and alert.
- A hunched appearance. A healthy chicken has an upright gait and will not be hunched over.
- Wounds on the legs. The skin on the legs of the chicken should be free of wounds.
- Bald spots without feathers usually indicate that the chicken may have mites or lice.
- Crooked beak.
- Coughing or wheezing are signs that the chicken is sick.
- A droopy head is a sign of illness.

What to look out for when buying chickens for laying eggs:

If you are specifically looking to buy good egg-layers, then there are a few pointers that will help you identify chickens that have already started laying eggs.

- Check for bright combs and bright eyes. If a pullet has a dull comb, they have probably not reached egg-laying age.
- Chickens that are have already started laying eggs have wide hip bones as opposed to the narrow hips found in pullets that have not yet reached the egg-laying stage.

How Much Should You Expect to Pay?

Chicken prices will generally vary from breeder to breeder. Some breeds also tend to cost more than others, so all this will influence how much you will pay for the chickens you want. However, here are some guidelines with average indicative costs.

I. Chicks tend to be cheaper than mature birds. Chicks for most breeds cost anywhere between a dollar and five dollars.

II. Pullets that range in age from 1 month to 4 months (4 -16 weeks) will, on average, cost between $15 and $ 25.

III. Mature chickens or laying hens will range anywhere from $10 to $100 depending on the breed.

Baby Chicks or Hens – Which are Better?

When starting your backyard flock, you may be wondering whether to go for mature chickens or start with chicks. The choice will really come down to whether you are willing to wait the six months it takes for the chicks to mature and start laying eggs. Some people choose to go for mature chickens because they do not require as much care as baby chicks do. Ultimately, make a choice based on your circumstances and how well-equipped you are to take care of baby chicks.

Baby Chicks Pros
- Cheaper than mature chickens
- Require less feed
- Easier bonding with your pet

Cons
- Six months wait for eggs
- Require more attention and care

Buying pullets, which are adolescent hens of 15-22 weeks of age, may ultimately be better for you if your sole purpose for raising chicken is eggs. This is because pullets are usually just about to start laying eggs and will give you more than older hens.

If you do go for chicks, then you will need to have a brooder ready for them. A brooder is basically the first place where your chicks will go when they get home. It helps to keep them warm and well-insulated at their tender age. You do not have to buy a brooder; you can improvise using a container or a cardboard box. Just make sure that your brooder, whether store-bought or improvised, has at least 2 square feet of space for every chick.

It is important to make sure that your brooder is deep enough –at least 12 inches deep. This keeps your chicks safe and ensures they do not jump over the sides. You do not need to cover the brooder if it is deep enough. However, if you choose to cover it, make sure you use a breathable material since your chicks will need ventilation.

To maintain the temperature in the brooder at the required level, you will need a brooder lamp. You can buy a 250-watt heat lamp from most hardware stores or feed stores. The lamp should be mounted with a clamp to avoid fire hazards. Finally, you will need to put bedding on the brooder floor to keep the brooder moisture free and well insulated. Pine shavings are recommended as an ideal bedding material for brooders.

Once you have set up the brooder, you can now put in your chick feeder and waterer. Buy a feeder that is specifically designed for chicks since it will be easier for your chicks to use. It is best to have your brooder set up before the chicks arrive. Chicks tend to be delicate, and making sure that their brooder is well set up and ready for them will help you get off on the right foot.

Your chicks will stay in the brooder for about five weeks. After this period, they can be safely moved to the main chicken coop. You may want to keep them inside the coop for the first couple of days so that they can understand that the coop is their "home." Once they get used to the coop, they can be allowed some time outdoors in the chicken run.

Chapter 7: How to Feed and Water Your Flock

A healthy flock is a happy flock, and this will only be achieved if your chickens are fed on a wholesome and balanced diet. Chickens, for the most part, are not fussy eaters and will happily eat scraps from your table, bugs, and weeds on the ground and, of course, chicken feed. This means that feeding your chickens is not going to be too much of a hassle, provided you know what a healthy chicken diet consists of.

Feeding Chickens at Different Life Stages

The nutritional needs of a chick will vary from those of a mature hen or even a pullet. This means you need to feed your chicken's age-appropriate feed that will best meet the nutritional requirements for the stage of life they are in.

Starter Feed

Baby chicks, ranging from day-old chicks to 18 weeks old, require a starter diet. At this tender age, chicks need plenty of protein to promote growth and development. This is why starter feed is recommended for chicks as it contains more protein than any other type of chicken feed. On average, starter feed for baby chicks will have a protein content of 22%. This protein is essential for healthy growth and for the formation of feathers, which are predominantly constituted of protein.

Another reason you need to make sure that you are only feeding your chicks starter feed is that it contains low levels of calcium. Chicks are super sensitive to calcium, and if they consume high amounts of this mineral, it can lead to deformities in the bones and even cause kidney damage. Feed for layers, or mature hens, tends to have high levels of calcium, which is required for egg formation. While this is beneficial to the egg-laying hens, for baby chicks too much calcium is detrimental, and you should therefore never feed your chicks layer feed even for a day.

Your small chicks will have tiny beaks, so they require feed that is ground into fine pieces to make it easier for them to eat and digest. Starter feed is designed to be fine enough for baby chicks, so while your chicks are below 18 weeks of age, the best feed for them is starter feed.

When shopping for starter feed, you will notice that starter feed is available in both medicated and non-medicated options. If your chicks have been vaccinated against coccidiosis, do not feed them medicated starter feed. The medicated chicken feed contains amprolium, a compound that tends to affect the efficacy of the vaccine. On the other hand, if your chicks have not been vaccinated against coccidiosis, the amprolium in medicated feed serves to protect your chicks from the disease.

Keeping your chicks in hygienic conditions helps to boost their natural immunity, so you do not necessarily need to feed your chicks on medicated starter feed to keep them safe from diseases.

Grower Feed

From eight weeks old, your chicks need grower feed, which is designed to keep your chicks growing until they reach laying age at between 18 t0 22 weeks. Since chicks in the growing stage, between 8 to 18 weeks, are not yet laying eggs, grower feed contains less calcium than layer feed. The protein content in grower feed is not quite as high as that in starter feed, but it is sufficient to help your chicks mature into layers.

Just like baby chicks, pullets need age-appropriate food because it meets their nutritional needs best. Do not feed your pullets or growing chicks layer feed because it contains too much calcium for that age and may cause health issues in the long run.

Layer Feed

Layer feed is appropriate for chickens over 18 weeks old that have started laying eggs. Layer feed is designed to provide all the essential nutrients needed to keep your mature chickens healthy and productive in terms of egg-laying. Layer feed contains more calcium than either the grower or starter feeds, and this is because laying hens need more calcium for proper shell formation.

The protein content in the layer field is at about 16%, and while this is sufficient to meet the nutritional needs of a mature chicken, it would be too little to meet the needs of baby chicks or growing pullets. That is why it is important to only feed layer feed to adult hens of 18 plus weeks.

Broiler Feed

If you are raising your chickens for meat purposes, the recommended feed for them is broiler feed. This type of feed is rich in protein and is formulated to promote faster growth and boost weight gain. It helps your chicken to put on weight fast, which is a desirable characteristic in birds bred for meat purposes.

Do not feed broiler feed to layers since it does not have the necessary nutrient content to boost egg production.

Different Forms of Chicken Feed

Chicken feeds are available in different forms. They can be in the form of mash, pellets, or crumble. Mash is a fine loose chicken feed that is easy to digest. You will find that chick feed comes in the form of mash since this is the easiest form for small chicks to digest. Grower feed also often comes in mash form as does layer feed.

Crumble is a semi-loose type of chicken food. It is coarser than mash and can be fed to either pullets or layers. Apart from crumble, chicken feed is also available in the form of pellets. Pellets are popular since they tend to be less messy compared to mash, and most people find them to be easier to handle.

Supplements

• Crushed Oyster Shells

Laying hens may need an additional source of calcium in addition to what is in their feed. That is why crushed oyster shells are recommended for laying birds. The supplemental amount of calcium helps to boost egg production and shell formation. The oyster shells do not need to be mixed with the chicken feed, simply provide them in a separate trough or feeder.

Chickens can control their calcium intake based on what they need, so you do not need to worry about providing them with too much shell grit. They will only eat what their body needs.

• Grit

Grit is used to refer to hard materials such as sand, small stones, or dirt that are provided to chicken to aid in digestion. Chickens need grit in their diet to enable them to digest fibrous foods such as grains in their gizzard. If your chickens spend time outdoors, they will pick up grit from the ground as they scratch and dig around so they will not need additional grit in their diet. However, if your chickens are confined to their coop, you need to provide them with some grit in a separate container to enable them to digest fibrous foods.

Chicks and pullets that are only fed on starter feed or growers feed do not require any grit since they are not fed grains or other foods that are hard to digest.

• Treats and Scraps

Human food is generally safe for chickens, and there is no problem feeding chickens scraps from your table. However, since they are getting all their nutritional requirements met by their chicken feed, it is recommended that you keep treats and table scraps to a minimum. Avoid feeding your chickens fatty foods as this may lead to obesity and, in some cases, even hinder egg production.

Treats such as chicken scratch can be used to boost your chickens' carbohydrate intake. Chicken scratch typically contains a mixture of different grains. Though grains are good for your chicken, they do not contain all the nutrients that your birds need, so always use chicken scratch as a treat and not the main staple of your chickens' diet.

Too much chicken scratch can lead to obesity since it is high in carbohydrates. Chicken scratch also lacks all the nutrients your chickens need, so relying on it as your primary chicken feed is not recommended. As long as you are feeding your chickens the appropriate feed, treats and scraps may be given occasionally, but they are not necessary.

Free Feeding vs. Restricted Feeding

Chickens eat pretty much all day, so restricted feeding is not really recommended. A chicken can only eat small bits of feed at a time, so in most cases, they will eat a little at a time throughout the day. By using feeders that replenish feed as it is eaten, you can ensure that your chickens have access to feed through the day without necessarily having to keep replenishing their feed manually.

Finding the right feeder for your chickens will make the feeding process so much easier. Feeding chickens on the ground is not recommended, as when this is done, the feed ends up mixing with chicken poop and other kinds of dirt in the ground. This can lead to diseases and infections in your flock. To avoid this, an appropriate feeder will come in handy.

Automatic Feeders

Automatic feeders are convenient, easy to use, and also help to reduce the waste of feed. With this kind of feeder, you will store your chicken feed in it, eliminating the need to keep refilling your feeder.

With an automatic feeder, the feed is dispensed as needed, so you may end up saving on chicken feed costs. The downside to this is that since chickens can access the feed anytime, it can encourage overeating, so there are both pros and cons to using automatic feeders. These types of feeders are also effective in keeping pests and bugs away from the chicken feed. However, an automatic feeder is typically more expensive than other types of feeders.

Ultimately, if you do not fancy having to feed your chickens manually every other day, an automatic feeder is the way to go.

Gravity Feeders

These feeders are simple to use and operate by simply releasing feed downward as it is eaten. You can mount a gravity feeder or leave it as free-standing, depending on where you choose to position it. Unlike the automatic feeder, this type needs to be replenished often since you can only put in a limited amount of feed.

The number of feeders that you will need will depend on the size of your flock. Aim to have at least one feeder for every ten birds. If you have birds of mixed ages in the same coop, you should have a separate feeder for your chicks to ensure that they are only eating their feed and not layer feed.

Gravity feeders are inexpensive, easy to use, and a convenient option if you have a small flock.

Watering your Chickens

Mature chickens drink approximately a pint of water daily. Since they drink this amount in small portions throughout the day, it is essential to make sure that your chickens have full-time access to clean water. Lack of enough drinking water can cause poor egg production, ill health, and even poor development.

Waterers come in handy because they help to deliver water efficiently to your chickens. When you provide water to your chickens in open containers, chances of dirt and debris contaminating the water is high. This is why a waterer is more suitable and hygienic for your flock.

Galvanized Waterers

When using a galvanized waterer, vacuum pressure allows water to keep filling the drinking trough as needed. This limits waste and prevents overfilling. However, you will need to put the waterer on a level surface for it to function properly. Alternatively, you can also suspend or hang your waterer from the roof of the coop.

A galvanized waterer is typically made out of steel and is therefore very durable. However, if you plan to supplement the water with vinegar or other supplements, they will react with the metal, so it is best if you go for a plastic waterer.

Plastic Waterers

Just like galvanized waterers, plastic waterers release water as it is needed. This helps to eliminate waste and also keep the drinking water clean. Plastic waterers come in a variety of sizes ranging from small chick-appropriate sizes to larger ones. This type of waterer is easy to use and is the most popular type of waterer among people who raise backyard chickens.

With this kind of waterer, you can add supplements to the water since the supplements will not react with the plastic. This type of waterer is also great for extreme heat conditions since it will not heat up as fast as galvanized metal. Plastic waterers also insulate the water better than metal waterers in cold temperatures.

Nipple Waterers

Nipple waterers typically have little plastic nipples or outlets attached to the main waterer so that instead of drinking from a trough or lip, your chickens drink from the nipple. These waterers help to keep the mess to a minimum. However, you will need to train your chickens to drink from this type of waterer until they get the hang of it.

Some waterers will have cups instead of nipple outlets. These can be bought separately to attach to your normal waterer, or you can buy a waterer that already has them attached.

Homemade Waterers

You can easily fashion a homemade waterer for your chickens using a plastic bucket and a dish. Simply drill some holes on the bucket. Drill the holes lower than the top of the plastic dish you will be using. Fill the bucket with water and replace the lid. Your bucket should sit on top of the plate allowing a drinking area for the chickens along the edges.

Watering in Winter

When temperatures drop, water tends to freeze, so you will need to make sure that your chickens still have access to water during winter. Replenishing your waterers with warm water often is one way to ensure that your chickens have access to drinking water during the cold season.

If you are using a galvanized waterer, having a heat lamp directly over it can help to prevent the water from freezing. Some waterers come with heated bases that can be connected to electricity to keep the water from freezing. This will also help in ensuring that your chickens have access to water during the cold winter months.

Signs of Poor Nutrition in Chicken

For your chickens to remain healthy and productive, a healthy diet is crucial. Taking care to observe any signs of nutritional deficiencies in your flock will guide you in knowing whether you are feeding your chickens properly.

Here are some classic symptoms of poor nutrition that you need to be on the lookout for.

I. A drop in egg production

II. Poor feathering

III. Eggs with thin shells

IV. Curved legs

V. Stunted growth

VI. Ruffled feathers

VII. Toes that curl inwards

VIII. Chickens eating their own eggs

Proteins, vitamins, minerals, and carbohydrates all play a crucial role in ensuring good health in chickens. Always ensure that your chickens are getting the right feed that is age-appropriate. If your chickens are enclosed or restricted to an area where they cannot forage on the ground, you can include supplements in their diet to make up for any nutrient deficiencies in their food.

Ultimately you will only get the best out of your chicken if it is well fed and cared for. Healthy birds produce more eggs than those that are in poor health. Always check labels of the feeds you are buying to ensure that you are getting feed that meets the basic nutritional requirements of your chickens.

Chapter 8: Managing Your Laying Hens

With the increase in awareness of the importance of healthy food from healthy food sources, it is no surprise that an increasing number of people have turned to backyard chicken farming. Whether you keep a modest flock or dozens of birds, one thing most people agree on is that having your own supply of fresh eggs is pretty convenient. However, to get the best out of your layers, you need to ensure that they are well taken care of.

Start with Healthy Chicks

If you are raising backyard chickens for the eggs, you can either buy mature hens or chicks. Chicks tend to be cheaper to buy compared to adult hens. However, there will be a waiting period before you can start collecting eggs. Chicks may mean a lot more work in terms of care and maintenance, but once they start laying eggs, they are likely to produce more eggs than adult hens. On the other hand, chicks do require a lot of care, so if you do not have a lot of time for care and maintenance, you can always buy adult hens.

If you choose to start your flock with chicks, the kind of care they get at the early stages of life will definitely impact their egg production in adulthood. One common mistake you need to avoid is feeding baby chicks food for layers. Even if your chicks are meant to be bred into egg-layers, they should never be fed layer feed until they are at least 18 weeks old.

Layer feed has high levels of calcium that is beneficial for egg-laying hens since it helps in shell formation. However, chicks do not require high calcium levels, and if they consume too much of it, it can lead to kidney problems and bone deformities. Always feed your chicks on starter feed until they are 18 weeks old. After 18 weeks, most breeds will be ready to start laying eggs, and at this point, you can safely switch them from starter feed to layer feed.

Your chicks should always have access to clean water. You can have your chick waterer suspended above the floor of the brooder so that it does not get contaminated by poop or any other dirt. You must always keep the brooder clean if you want them to stay healthy. If you leave the bedding in your brooder too long, the accumulation of poop and moisture can lead to diseases.

A dirty brooder can undermine all your hard work even if you are feeding your chicks the right feed. Clean your brooder as often as possible, and do not let the bedding get moist. A dirty brooder can lead to diseases and negatively impact the proper growth and development of your chicks.

Once the chicks have started to grow some feathers, usually in about five-to-six weeks, they are ready to go to the main coop. From this age on, they can be allowed outdoors, although you will need to make sure that they are safe from predators and rodents.

Feeding your Layers

When your chickens reach the egg-laying stage, their nutritional needs evolve to enable them to produce eggs. This means that they need to be fed layer feed. This feed has a healthy dose of calcium, which is required for shell formation. It is important to ensure that your layers are eating the right food and that they have enough of it.

Chickens tend to eat throughout the day. The best way to feed them is through feeders that dispense food as it is eaten. This helps to ensure that your layers have access to food whenever they need it. The primary source of nutrition for your layers should be layer feed. While chickens are happy to eat anything, including scraps from your table, they do require a nutritional diet, which is only achieved by feeding them mostly on high-quality layer feed.

Even with high-quality layer feed, your layers will still require an additional source of calcium. That is why it is important to provide your chickens with crushed oyster shells. Crushed oyster shells are a great source of calcium for layers. All you need to do is put them in a separate dish or container when feeding your chickens. You do not need to worry about your chickens eating too much of the oyster shells. Chickens will eat just as much calcium as their body requires. Make oyster shells a staple in your chicken's diet if you want to boost proper egg production.

If your chickens are not free-range, it means they may not be getting enough grit. Grit is coarse material that chickens ingest from the ground to aid in the digestion of fibrous materials such as grains. For enclosed birds, you will need to provide them with grit to complement their diet and aid in proper digestion. Do not mix the grit with the regular chicken feed but rather provide it separately. Just like the oyster shells, chickens will only ingest as much grit as they require, so you do not need to worry that they may eat too much of it.

Apart from their main layer feed, here are some treats that you can include in your chickens' diet to keep them laying eggs.

- **Mealworms**

This treat is full of healthy protein and makes a very healthy treat for chickens. It also contains plenty of essential minerals and vitamins that are good for your chickens' health. However, do not feed too much of it to your layers as they do not require excessive amounts of protein. A spoonful of mealworms per chicken once or twice a week should be sufficient.

- **Cracked Corn**

This is a healthy treat for layers. Corn is, however, high in carbohydrates, so it should be fed to chickens in moderation to avoid obesity. Excess weight gain reduces egg production and is not good for your chickens' health.

- **Greens**

Leafy greens and vegetables have plenty of essential minerals and vitamins that help to keep your chickens healthy. Kale, cabbage, and dandelions all make great healthy treats that you can feed to your chickens occasionally.

Fruits like watermelon are also good for your chickens and can be given as treats every so often.

- **Scratch Grains**

You can feed your chickens scratch grains as treats provided you do so in moderation.

- **Scraps and Leftovers**

Chickens can safely consume human food, as most human food is also safe for chickens. However, when giving your chickens scraps from your table, avoid certain foods, including avocado, tomato stems, and fruits like lemon and oranges. Foods like garlic and onions should also be avoided. Bear in mind that table scraps should be given in moderation as they can lead to obesity, which in turn will affect the health of your chickens.

Finally, your layers need to have access to clean water at all times. Find a suitable waterer to use and ensure that you always keep it clean. Contaminants from chicken poop, debt, and debris can easily contaminate water. If you find that the chicken water has dirt in it, pour it out and replace it with clean drinking water.

Housing

Your chicken coop and chicken run need to be kept clean to ensure the health of your chickens. Make sure you have suitable bedding to keep the coop moisture-free. Bedding also helps to prevent the accumulation of ammonia in the coop. If ammonia from chicken manure builds up to very high levels, it can lead to respiratory diseases, so it is best to ensure that your chicken coop is well ventilated.

Layers need a private space for laying eggs and for brooding. You should have nest boxes in your coop where your layers can lay eggs. The nest boxes need to be cushioned with bedding. Straw and hay make great bedding for nest boxes since they are soft. This bedding will cushion the egg once they are laid and also help to insulate the chicken. However, just like the bedding in the rest of the coop, change the bedding in the nest boxes often to keep them clean. Nest boxes should be cleaned at least once a month.

Nest boxes need to be slightly raised off the ground. Your nest boxes should be dimly lit, so just ensure that they are not positioned in an area with direct sunlight. Some people use curtains, but this is not necessary provided you have positioned the nest boxes in a quiet area of the coop.

You need at least one nest box per four chickens so that your layers all have access to a next box when they need to lay eggs. If the nest boxes are too few, your chickens may resort to laying eggs in hidden nooks and crannies that may be hard for you to reach. Most of all, make sure that your nest boxes are safe from predators, rodents, and other pests. Eggs can attract predators, so your nest boxes need to be checked regularly for pests and rodents such as mice.

Winter months pose a challenge for chickens since the cold weather can affect egg production if your flock is not comfortable and properly insulated. To make sure that your layers are comfortable during the cold season, here are the factors you need to keep in mind.

1) Light

Layers need light to stimulate the pineal gland. This gland initiates egg production by releasing hormones to initiate the process. This means that your chickens need daylight to produce eggs. In winter months, you can substitute daylight with a 60-watt incandescent light bulb. Make sure you provide light for at least 16 hours each day.

2) Roost

Roosting bars are an essential part of any coop. Your layers need a comfortable area to perch on, and roosts provide them with this space. When it is cold, chickens tend to roost close together to keep each other warm. Ensure that coop has sufficient roosting space for your layers. A general rule of thumb is to have at least 8 inches of roost space per chicken.

3) Keep the Water Supply from Freezing

Water tends to freeze in winter, especially if it is in galvanized waterers. This means you will need to keep a fresh supply of warm water to your layers during colder months. Chickens will not lay eggs if they do not have access to sufficient water, so making sure that their water is not frozen over is crucial.

4) Deep Litter Method

Bedding does not just help to keep the coop clean and odor-free; it also helps to insulate the coop, making it warmer and more comfortable for your chickens. During winter having a deeper layer of bedding and litter than you do during summer months can help to keep the coop warm.

To use the deep litter method to keep the chicken coop warm enough in winter, you can simply keep adding on to your regular bedding as winter approaches by adding layer after layer of bedding periodically. By winter, if your litter is up to 8 inches deep, the lower layers of bedding will start giving off heat as they compost, and the coop will be much warmer.

5) Provide Warming Treats

Treats such as corn, which boost metabolism in chickens, can help to keep the chickens warm. You can feed cracked corn to your chickens in the evenings to keep them warm during the night.

Reduce Stress for Better Egg Production

Winds, extreme heat, and cold snaps are all stress factors that can affect your chickens' ability to produce eggs. To reduce the levels of stress that your chickens go through during harsh conditions, here are some easy tips to help keep egg production up and your chickens healthy.

● Boosting protein intake can help to minimize the effects of stress on your chickens. You can add protein-rich treats such as mealworms to your chickens' diet.

● Green feeds such as vegetables can help to boost fertility and egg production in chickens. They are rich in essential vitamins and minerals and will have beneficial effects, especially during high-stress seasons.

● Adding vitamins or supplements to drinking water may also help in boosting egg production.

● Heat stress is also bad for chickens and can cause a decrease in egg production. Ensure that during super-hot months your chickens have access to shaded areas where they can cool off.

Reasons Why Chickens Stop Laying Eggs

1. A poor diet is one of the primary reasons why chickens stop laying eggs. Always feed your layers the recommended feed for egg-laying chickens and include healthy treats such as oyster shells to boost calcium levels in the diet.

2. If your chickens are not getting enough daylight, they may stop laying eggs. Chickens need at least 16 hours of daylight to produce eggs. An artificial light source can help you ensure that your chickens get sufficient hours of light every day.

3. Brooding hens do not lay eggs. They will typically spend a lot of time in the nest box and may become protective over their space since they are trying to hatch eggs. This process usually lasts for 21 days.

4. Some chicken breeds are not as prolific layers as others, and may only lay two or three eggs in a week.

5. Diseases and parasitic infections can interfere with egg production, so if your chickens are in ill-health, the chances are that their egg production will drop.

6. Chickens will eventually stop laying eggs due to old age. Most chickens will actively lay eggs for about three years, but after that, there will be a natural decrease in egg production until it ceases altogether.

Chapter 9: Understanding Chickens

As a first-time owner, you may observe behavior in your chickens that you may not understand. Chickens tend to vary in terms of temperament, preferences, quirks, and even activity levels. Your chickens will have different personalities, and so it is not unusual to find that your flock is composed of chickens that each have their unique characteristics. Understanding why your chickens behave a certain way may help you in caring for them and bonding with them.

Chickens do better when raised in groups or flocks. This is because chickens are naturally social, so they thrive in groups or flocks where they are part of a family or community. Watching your chickens as they interact and go about their business can be quite interesting, and many people who enjoy raising chickens as a hobby often find themselves enjoying "chicken TV". However, it is important to know what behaviors constitute normal chicken behavior and which may be signs of illness or stress.

Normal Chicken Behavior

Pecking Order

In every flock, there is a pecking order. This is just the way the social hierarchy of chickens works. Whenever chickens are put together in groups, they will find ways to establish ranks where there is a kind of social structure, and everybody knows their place. This happens even amongst baby chicks, and you will find that even birds at that tender age have a pecking order.

Chickens will often fight each other to establish and maintain a pecking order, so do not be surprised to see the chickens in your flock engage in squabbles from time to time. Most of the fights are usually short-lived and do not really end up causing grievous harm. This, however, may not be the case if your flock has several roosters. Roosters can fight to the death, especially if there are hens in the flock to fight over.

Small flocks will tend to have fewer squabbles over rank for the simple reason that it is easier to establish a pecking order in smaller groups. Larger flocks will have more frequent squabbles, especially if the flock has several roosters. In any flock where there is only one rooster, he will dominate the hens and will be the default leader of the flock.

The lead rooster maintains the social hierarchy in his flock and even comes to the rescue when the hens in his flock are squabbling. Although the rooster becomes protective of all of the hens in his flock, often, he will have a favorite hen whom he obviously favors over the others. This kind of favoritism is a mating ritual of sorts, and the rooster will mate more often with his favorite hen than with the other hens in his flock.

In flocks without a rooster, a dominant hen becomes the leader of the flock. She will be in charge of the flock and will take on the role of protector and peacemaker for the rest of the flock. The pecking order is usually disrupted when new chickens are brought into the flock. If you introduce new birds to an existing flock, there is likely to be a certain amount of squabbling. This is usually a way to show the new birds who's in charge and establish their places or rankings in the flock.

To keep squabbles to a minimum when you bring new birds home, you can separate them from the rest of the flock for a day or two. This will give them time to get familiar with each other without necessarily being within fighting distance.

Ultimately, squabbles and fights between chickens are normal. It is simply how they establish their pecking order. You do not need to try and separate them during such fights since these squabbles are usually short, and in most cases, no blood is shed. However, roosters can kill each other in the course of serious fighting, so once they start to draw blood, you should separate them to avoid a fatal outcome.

Dust Bathing

You will often notice that your chickens love bathing themselves in the dust. Dust bathing is a normal behavior in chickens, and providing a dust bath in your chicken run is actually recommended in order to keep your chickens happy. When taking a dust bath, your chicken will find a spot with loose soil. They will then dig a depression in this patch before sitting in it and throwing the soil over themselves with their feathers and legs.

Dust bathing helps chickens to get rid of mites, lice, and other parasites. It is also a fun experience for them, so if your chickens are confined to a run, you can always provide them with a box of sand for dust bathing.

Broodiness

Hens will get broody every so often. This is when the hen wants to hatch eggs. Brooding hens become inactive and may sit in the nest box for days on end. They may also become aggressive or protective as they try to guard their nesting space. If you want chicks, this is the best time to put eggs in your chicken's nest box and let her hatch them. However, if you do not want chicks, you can stop a chicken from being broody by lowering their body temperature. A cold bath or keeping them from the nest at night are two easy ways to do this.

Crowing

Roosters crow every day. This is just part of their nature; they will crow at the crack of dawn and throughout the day. This is part of the reason that most cities and towns ban the keeping of roosters. The crowing is, unfortunately, not something you can stop, since it is a natural part of a rooster's behavior.

When crowing, roosters are essentially making their presence known to the other roosters, to the hens around them, or simply just expressing themselves. Even if your area allows you to raise roosters in your backyard, be sure that you are ready to contend with them crowing because it is going to be an everyday occurrence.

Preening

You may notice that your chickens spend quite a bit of time pecking at their feathers. This kind of preening is part of normal chicken behavior. When preening, chickens are essentially removing dirt, pests, or bugs from their feathers, so in effect, preening is your chicken's way of keeping themselves clean.

Molting

Chickens usually undergo a period where they shed old feathers and grow new ones. This process is referred to as molting and will typically occur when temperatures start getting cooler. During this molting period, you will notice that your chickens will stop laying eggs as they are reserving their nutrients for the feather renewal process.

The molting period tends to vary from chicken to chicken, but on average, it will range from 4 to 16 weeks. You can help to speed the process up for your chickens by boosting their protein intake. Feathers are mostly made up of protein, so the higher the amount of protein in your chicken's diet, the faster the molting process will be. It is also best to avoid stressing your chickens during this period. This means you should not handle or touch them as their bodies are super-sensitive during the molting period.

Scratching and Digging

Chickens love to scratch and dig the ground. They dig up bugs, worms, and grit to eat from the ground. You will notice that your chickens will spend most of their time outdoors digging and scratching. This is normal and expected chicken behavior, and it is recommended that you provide your chickens with a foraging area outdoors where they can dig and scratch. If your chickens are not free-range, you can confine them to a run, which will give them foraging space where they can scratch and dig.

Celibacy

Hens do not need a rooster in the flock to be happy or to lay eggs. Hens that are raised without roosters create their own society and pecking order where the dominant hen becomes the head of the group. Hens will lay eggs as they normally would without a rooster in the flock. However, since these eggs will not be fertilized, they will not be able to hatch chicks from them.

Abnormal Chicken Behavior

There are chicken behaviors that should serve as an indicator that there is a problem with your flock. Abnormal chicken behavior can be caused by illness or stress factors, so observing your chickens often will help you pick up on any unwanted behavior. Here are some abnormal chicken behaviors for which you need to be on the lookout.

Aggression

Squabbles and fighting, as we have learned, are acceptable and normal behaviors that chickens use to establish a pecking order in the flock. However, in some cases, you may have overly aggressive birds that attack you or your children. This behavior is commonly observed in roosters. Roosters may get into a habit of charging anyone who comes near their space. They will peck, slap with their feathers, and try to hit with their claws or spurs.

This kind of aggression can be dangerous, especially if you have children, and needs to be addressed. A rooster that attacks humans is trying to establish dominance over them, and if the behavior is not curbed, it can become a serious problem. To establish that you are boss, you need to handle the rooster a bit forcefully. This means that if he pecks at your feet, you should shove him with your feet and force him directly to the ground.

The aim here is not to hurt the bird but rather to force it into a submissive position. You can also hold him down for a few minutes. Forcing the aggressive rooster to stay still is a way of establishing who is in charge. Roosters will generally stop attacking humans once they are made to understand that humans rank higher in the pecking order.

Hens will rarely be aggressive toward humans unless they are protecting their baby chicks. This is just a natural protective instinct and will only last while the chicks are young. Avoid touching or handling baby chicks as this may cause the mother hen to feel that her babies are in danger and attack.

Pecking and Feather Picking

When your chickens do not have enough space and are in an overcrowded coop or run, they may resort to incessant pecking and feather picking. This can be a sign of stress or boredom. Always ensure that your run and coop have enough space for the number of birds you have in your flock. The recommended space in the coop per chicken is at least three square feet while your chicken run should have at least eight to ten square feet of space for every bird.

Wing Dropping

If you observe your chickens dragging their feathers on the ground, this is a common sign of illness. Droopy wings can point to any number of conditions, and you may need to have your chicken checked for diseases. Remember that chicken diseases can spread pretty fast among the flock, so early intervention can help you prevent a disease from spreading.

Lethargy

A normal chicken is by nature alert, active, and curious. They will be scratching and digging at the ground, moving around constantly, and interacting in one way or another with the rest of the flock. If your chicken appears dull or inactive and exhibits general disinterest in what is going on around them, they may be unwell. If you observe that the chicken is having trouble holding its head up or walking, this is a clear sign of ill-health, and you need to have a vet look at it.

Hens Eating Their Own Eggs

When hens start eating their own eggs, this can become a serious problem. This behavior is commonly caused by a calcium deficiency, and the chickens start to eat eggs as a way to supplement their calcium intake. If you let this behavior go on for a long time, it will become even harder to break, so it is best to stop it as soon as you realize that it is happening.

To stop this habit, feed your chickens extra calcium by providing them with crushed oyster shells. Avoid feeding them their eggshells unless they are completely crushed; otherwise, they will start to associate their eggs with the shells you feed them.

Another reason that may encourage hens to eat their eggs is egg breakage. Once an egg breaks, then the hens will be likely to eat it. To avoid this, make sure that your nesting boxes are well cushioned with straw and hay. You also need to avoid congestion in the nesting boxes by having at least one nest box for every four chickens in your flock.

Ultimately chickens have unique quirks and personalities, and the best way to understand your flock is to spend some time watching them. In this way, you will get to know what the normal behavior for them is. Once you understand their routine behavior, it will become easier to identify any uncharacteristic behavior that may be caused by stress, illness, or other factors. A good chicken farmer knows their flock well and makes it their business to understand their chickens.

Chapter 10: All About Eggs

One of the main benefits of raising chickens in your backyard is the fresh supply of eggs. As more and more people look for ways to produce their own food and take control of what kind of food is on their table, the popularity of raising chickens keeps increasing. For beginners who have just started raising chickens, collecting your first batch of eggs can be quite satisfying.

Bright yolks, firm whites, and of course, tasty goodness are some of the hallmarks of fresh eggs. When you compare eggs from your coop with grocery-bought eggs, the difference is usually quite clear. When you purchase eggs from a grocery store, you have no way of knowing just how fresh they are, how the chickens that produced the eggs were raised, or what kind of feed they were given.

When you are raising chickens in your backyard, you have control over their diet. This means you can choose to go organic and, in this way, ensure that your eggs are completely natural and GMO-free. This is what makes raising backyard chickens so fulfilling. If you are a beginner, you will soon realize that eggs come in different shapes, colors, and even sizes. From brown eggs to white and even blue eggs, there is an array of not just colors but qualities of eggs.

Before you can get to enjoy your eggs, you first need to know the best practice when it comes to collecting cleaning and storing them.

Collecting Eggs

When your hens lay eggs, you do not want to leave them lying around in the coop for too long. Here are some of the reasons why collecting your eggs regularly is important,

- Eggs are fragile and the longer you leave them in the coop, the higher the chances of them getting trampled and broken.
- Eggs can attract predators and rodents to the coop. Cats, raccoons, rats, and other types of predators like the taste of eggs so they can get into the habit of getting into the chicken coop if eggs are constantly left lying about.
- Eggs do not have a very long shelf life, so if you want to enjoy your eggs fresh, it is best to collect them often from the coop.
- Hens can start to eat their own eggs if they are not collected often. This happens especially when there are broken eggs in the coop, and hens get into the habit of eating them.
- Coops tend to have plenty of contaminants in the form of chicken manure. You do not want your eggs to stay too long in the coop. The longer the eggs stay in the chicken coop, the more likely they are to get contaminated with dirt and chicken poop.

If you have a small flock, collecting eggs once in the morning and later in the evening is advisable. People with large flocks should collect eggs thrice a day. This will ensure that eggs laid during the day do not stay in the coop overnight. A plastic container should be sufficient when collecting eggs. Just be sure not to stack them too high to avoid accidental breakage.

Cleaning Eggs

When chickens lay eggs, they usually have a natural protective layer on them to keep them germ-free. However, it is normal for eggs to get some dirt on them in the coop, so cleaning them before storage is good practice. In most cases, it is recommended that you clean your eggs with a dry piece of cloth. Using a dry cloth will help to get the eggs clean without damaging their natural protective outer layer.

Alternatively, there are times when your eggs may have poop stains and other kinds of dirt that needs to be washed out. In such cases, it is okay to wet clean the eggs with some water. Ideally, when using wet cleaning, you should use warm water. Once the egg is clean, you can dry it with a paper towel and then place it on a rack.

Always ensure that the nest boxes and coop are kept clean, as this will reduce the chances of collecting dirty eggs. Clean the bedding in the nest boxes as often as possible, and this will give your hens a clean place to lay their eggs. Ultimately this means cleaner eggs for you.

Storing Your Eggs

Whether your eggs are simply for domestic consumption or sale, proper storage is important in preserving freshness. Once your eggs are clean, they should be stored in an egg carton. It is recommended that you indicate the date of collection for the eggs on the carton so that you know which eggs are the freshest. This is especially important if you collect a lot of eggs from your flock daily. If you do not separate them by date, you risk some of them going stale on you.

Always use eggs in the order in which they were collected. This prevents situations where some eggs go bad because they have been stored for too long. As a general rule, store your eggs in the refrigerator. Refrigerated eggs will, on average, have a shelf life of one month from the date of collection. Eggs that were not wet-cleaned after collection can last for several weeks stored at room temperature. Always wash your eggs before using them to get rid of any dirt or contaminants on the surface.

If you have stored your eggs for a while and are not sure whether they are still fresh, you can use a simple float test to find out. Fill a bowl with clean water, then place the egg inside the bowl; a fresh egg will sink to the bottom, while a stale egg will float on the water.

Determining Egg Quality

The quality of an egg is typically based on internal egg quality and external egg quality. External egg quality centers upon the external characteristics of the eggs, such as cleanliness, shape, and even texture. If you plan to sell your eggs, they need to be graded as A or AA. If your eggs are graded as B, they are not approved for sale in stores.

External quality starts with how clean your eggs are. Even though a chicken will lay an egg when it is nice and clean, eggs easily get dirty in the nesting box. This is why it is important to collect your eggs as often as possible to keep contamination to a minimum. You can always dry clean or wet clean your eggs to keep them clean though this will affect their shelf life.

The other aspect that affects the external quality grading of an egg is its shape. Eggs that are any shape other than oval are considered to be lower quality. This does not mean that their nutritional content is any lower; it simply indicates that their physical shape differs from the ideal oval shape of an egg. Similarly, eggs with rough or uneven shells are downgraded since they are more likely to break than those with smoother shells.

Interior quality is graded based on the quality of internal features of the egg, such as the yolk. When an egg is fresh, the egg yolk tends to be round and firm. However, as time passes, the yolk starts to absorb water from the egg white and increases in size. This means that the longer an egg is stored, the more its internal quality reduces.

The internal quality of an egg is not just affected by the passage of time but will be affected by a host of other factors. These include disease, temperature, humidity, and storage of the egg. This means that to get high-quality eggs, your hens need to be healthy and fed on a well-balanced diet. How you handle the eggs and store them may also cause the internal quality of the egg to deteriorate.

When eggs are stored in high temperatures, the internal quality of the egg is reduced. This is why refrigeration is recommended in order to keep the eggs fresh for as long as possible. Rough handling may also interfere with internal egg quality, so always be gentle when collecting, cleaning, and storing your eggs.

Enjoying Your Eggs

Eggs are some of the most versatile foods on the planet. From breakfast to dinner and even desserts, eggs are a staple in many homes. They are used to create a wide variety of dishes. Starting with your morning omelet, your favorite pastry, salad, and many other meals, you are likely to find that eggs are ingredients in many staple dishes in a lot of homes. This is what makes having your own supply of fresh eggs so rewarding. Every time you use an egg from your backyard chickens, you can be sure of the freshness and quality of that egg.

What exactly is in an egg, and what makes this superfood so popular across the globe? Let's have a look at the nutrient content of an egg (boiled egg, values per 100 grams)

- Total fat 11 g (16%)
- Saturated fat 3.3 g (16%)
- Polyunsaturated fat 1.4 g
- Monounsaturated fat 4.1 g
- Cholesterol 373 mg (124%)
- Sodium 124 mg (5%)
- Potassium 126 mg (3%)
- Total Carbohydrate 1.1 g (0%)
- Dietary fiber 0 g (0%)
- Sugar 1.1 g
- Protein 13 g (26%)
- Vitamin A (10%)
- Vitamin C 0%
- Calcium 5%
- Iron 6%
- Vitamin D 21%
- Vitamin B-6 5%
- Cobalamin 18%
- Magnesium 2%

The humble egg carries a host of essential vitamins and nutrients, including protein. Eggs are also relatively low in calories, and this means they go well even with calorie-restrictive diets. Eggs are, in fact, a popular menu item for keto-dieters, so you should not fret too much about gaining excess weight from eating eggs.

If you have a constant supply of eggs from your backyard flock, remember there is plenty you can do with eggs in the kitchen. Try exciting new recipes, use them for baking your baked treats, and pretty much use your eggs as creatively as possible. There are plenty of egg recipes available online, so if you are looking for new ways to enjoy your eggs, there is always a recipe that you can try out and enjoy.

Chapter 11: Meat Birds

An increasing number of people are raising chickens for meat purposes. This is because as people become more sensitized to harmful chicken farming practices, they are choosing to have more control over the kind of food they eat. Factory farmed chickens are often not given the best care or feed, and raising your own birds for meat will give you access to more wholesome chicken meat.

You can easily raise chickens for meat in your backyard as the process of raising meat birds is pretty much the same as what you would do with any other chicken. The only difference usually comes in the type of feed that you will give your chickens if you are raising them solely for meat purposes.

The Best Meat Chicken Breeds

Meat birds vary from layers in that they tend to grow faster and also put on more weight. This means that while any chicken can be raised for meat purposes, ultimately meat breeds will give you more meat in a much faster time frame than layers. Here are some of the best meat chicken breeds that should be part of your flock if you are raising backyard chickens for meat purposes.

- **Jumbo Cornish Cross**

This is a large chicken breed that puts on weight pretty fast. They have large breasts and big thighs that have made them popular among chicken meat breeders. In about eight weeks you can expect a male Jumbo Cornish to weigh about four pounds, while a female will weigh in at two pounds at the same age.

- **Cornish Roaster**

This is another large chicken breed that is ideal for meat purposes. It has yellow skin, and, like the Jumbo Cornish, large breasts and thick thighs. This chicken breed matures fast, reaching maturity in about ten weeks.

- **Jersey Giant**

As the name implies, this is a large bird that is a favorite for many chicken meat farmers. It also has good egg production so it can serve as both a layer and a meat bird in your flock. Jersey Giants do not mature as fast as other meat birds, but it will grow to a considerable size and weight.

- **Freedom Rangers**

Freedom Rangers is another breed that is perfect for those who want to raise chickens for meat production. It is a large breed and will, on average, take nine to eleven weeks to mature.

Caring for Meat Chickens

Housing

Meat birds tend to be larger than your average layer, so they will require plenty of space. You need to have adequate room both in your coop and in the chicken run for your meat birds. Most meat breeds will grow much faster than the laying birds, so this means space in the coop will free up every so often. However, always ensure that each bird has a minimum of 3 square feet of space in the coop, and if you have a run, the minimum space allowance per bird is at least 8 square feet.

Overcrowded birds tend to spread diseases amongst themselves, fight more, and generally experience more stress. Remember, a healthy bird will give you healthy meat, so even birds that are being raised for meat purposes need to be kept in a healthy and comfortable environment.

A build-up of ammonia in the coop and poor ventilation can also cause problems for your flock. Always ensure that your coop has enough air flowing in and vents to let the air out. Poorly ventilated coops are a breeding ground for disease, and the last thing you want is to get meat from a sick or infected chicken.

Hygiene is key in maintaining your meat birds in a healthy state. Ensure that there is bedding in the coop to help in keeping it clean. Go for bedding such as wood shavings that will absorb and release moisture quickly, leaving the coop dry and odor-free. Clean out the bedding at least monthly to avoid the build-up of manure, which can lead to high ammonia levels in the coop as well as breeding of pests and parasites in the chicken coop.

Meat birds need pretty much the same level of care and maintenance as layers do. Keep them in a clean, comfortable environment, and you will have fewer diseases, bird deaths, and behavioral problems to contend with.

Feeding Meat Birds

Just like any other chickens, your broiler chicks should be started on a starter feed. Starter feed is rich in protein and is specifically formulated to promote growth and proper development in chicks. Starter feed should be given to the baby chicks until they are three weeks old. From this point on, the chicks can be fed grower feed. This feed is meant to promote fast growth and weight gain.

Phase feeding allows your chickens to get all the nutrients they need for the particular age they are at, so it is always important to feed your meet birds age-appropriate food. The key advantage of raising your own meat birds is that you can choose whether to feed them organic or standard feeds. Organic feeds have similar formulations as standard feeds but are typically grown and processed under organic conditions that are certified and approved by relevant regulators. This means that when making organic feeds, companies cannot use anything treated with chemical fertilizers or pesticides, nor any genetically modified compounds. When you choose organic feeds for your meat birds, then you can be sure that the meat you will get once the bird is processed will be free of chemicals or GMO products.

Regardless of whether you choose standard or organic feed for your meat birds, always ensure that the feed you choose meets the basic nutritional requirements. Starter feed should contain at least 22% protein, while grower feeds should contain at least 18% protein. Avoid giving your meat birds layer feed since it contains less protein than broiler feed and may slow down the growth rate of your meat birds.

Once you have the right food, make sure that your meat birds are getting as much food as they need. Meat chickens will, on average, eat more than layers, but since they will mature faster, the average cost will not be that much higher. Have a feeder for every ten birds or so to ensure that each of your chickens has sufficient access to feed. If the feeders are not enough, smaller birds will be bullied and will not get enough food.

Chickens will, on average, drink more water than the feed they consume, so they always need to have access to clean water. You can use waterers in your coop or run to make sure that your meat birds stay hydrated and healthy. Always ensure that the water is clean and free of any dirt or debris.

Meat breeds will generally be ready to be processed at about ten weeks of age, though this can vary from breed to breed. Do not let your meat birds go unprocessed for too long. This is because since they put on a lot of weight very fast if you do not process them at the right time, and they can develop organ failure due to the excess weight they are carrying.

Safety

You are not fattening your meat birds for predators to feed on, so safety should be a top priority when raising meat birds. Your chicken coop should be well-secured and locked up safely at night to keep predators out. Foxes, raccoons, cats, and dogs are all partial to the taste of chicken, so if they get access to your chicken coop, disaster is sure to follow.

Air vents in the chicken coop should be covered with chicken mesh to ensure that only air can get in and out. Check the coop often for rodents, which can hide in the coop bedding and pose a threat to your flock. It is advisable to keep your feed in a different storage area. If you keep chicken feed in the coop, it may attract rodents and other pests to the chicken coop.

Your chicken runs should also be built with the safety of your birds in mind. The fencing should be done using chicken wire or other small mesh fencing material. This will help to keep predators away from your chickens. In some areas, hawks and owls can be a nuisance, so the chicken run may need a cover to stave off flying predators.

Always ensure that you do not let your other pets in the chicken area. Domestic cats and dogs can easily injure chickens, so it is always best to keep them away from your chicken coop and run.

Processing Chickens for Meat

Meat birds will generally be ready to be processed from about ten weeks of age, depending on the particular breed. Before you butcher your bird, make sure you have all the tools that you will need close at hand.

Tools required
- Knives – very sharp, with a 4-inch blade or longer
- Poultry killing cone – available from farm stores
- Bucket
- Clean water – you can run a garden hose to your butchering area
- Gloves
- Apron
- Tarp covered table
- Scalding water in a huge pot (enough to dunk your bird in)

- Paper towels
- Plastic bags or containers for storage

The Butchering Process

I. Once you have captured the bird you want to process, hold it upside down by its feet. In this position, the bird will pass out, making the butchering process easier.

II. Place the chicken in the killing cone

III. Holding the head firmly through the bottom of the killing cone, make a deep firm cut with a sharp knife on the throat.

IV. Once the throat is cut, let the blood drain into the bucket until it is completely drained.

V. Once the blood is drained, remove the killing cone and, still holding the bird upside down by its feet, dunk it into the scalding water.

VI. Make sure the water is hot enough to scald the skin. It should be at least 135F. Swirl the bird around in the scalding water until all the feathers are soaked in the water.

VII. You want the feathers to become loose, but you do not want the chicken skin to tear. Once you pull on a few feathers and they come away easily, remove the bird from the scalding water.

VIII. Hold the bird or suspend it over your bucket and start removing the feathers. You will make much faster progress if you rub your thumb and fingers against the grain of the feathers.

IX. Once you have removed the feathers, rinse the bird with clean water.

X. The next part is preparing the chicken for storage or use.

XI. Hang the chicken up by its feet.

XII. Make a cut from the chicken's groin downwards towards the neck area. As you make the cut, the internal organs will also flow downwards. Cut carefully so that you do not puncture the intestines or any of the other internal organs.

XIII. Once all the organs have fallen (or been pulled) out, rinse out the bird until the water runs clear.

XIV. Finally, you can place the clean bird on your tarp-covered worktable and prep it. You can quarter it by separating it at the joints, or you can store it whole until it is needed.

Chapter 12: Health Care and Maintenance

As far as pets go, chickens are pretty easy going and low maintenance. With a little effort and time on your side, you can have a healthy and happy backyard flock. Most cases of ill health, poor productivity, and death in chickens can be traced back to either a poor diet or unhygienic coop conditions. All this means is that with proper care and maintenance, you should be able to get the best out of your feathery pets.

When it comes to caring for your chickens, the best way to do it is to have scheduled tasks. In this way, nothing gets overlooked, and all your chickens' needs are met on time. So, for most people who raise backyard chickens, having tasks broken down into daily, weekly, and monthly maintenance tasks helps them to keep up with the care of their chickens. This approach will help you in saving time while still giving your pets the best care possible.

Daily Maintenance Tasks

Check the Waterer
Chickens need access to clean water all day every day to stay healthy. You may not need to refill your waterer daily, but you need

to ensure that the water is clean and that no dirt or debris has gotten into it. If the water is dirty, replace it with clean water.

Feed the Chickens

Feed your chickens daily. You can choose to use an automatic feeder or gravity feeder that dispenses feed as it is eaten. Always check your feeders daily to ensure that your chickens have enough feed.

Collect Eggs

Collect eggs daily. If you have a large flock of layers, you may need to do this twice or thrice a day to keep the eggs clean and avoid contamination. Leaving eggs in the coop for extended periods may attract predators, and, in some cases, it may cause chickens to start eating their own eggs.

Chicken Watching

Spend a few moments each day observing your flock. This will help you spot any abnormal behavior or signs of illness in your flock. This does not need to take up too much time, and even a few minutes a day can help you stay in touch.

Monthly Maintenance Tasks

Change the Bedding

The bedding in the chicken coop needs to be changed regularly to prevent the accumulation of manure. This is a task that can be done monthly to ensure that your chickens live in a clean and healthy environment. When bedding is not changed often, it can lead to infections and disease.

Clean the Nest Boxes

Just like the rest of the coop, nest boxes need to be kept clean. Remember that this is where your eggs will be laid, and you do not want them contaminated with chicken poop or other kinds of dirt. Change the bedding in the nest box monthly to keep it clean.

Clean your Waterers

At least once a month, make sure that the waterers have been deep-cleaned to remove any contaminants. You can use a mixture of bleach and water to disinfect them completely and then rinse them thoroughly with clean water. Water can easily become a carrier of pathogens and disease-causing germs, so keeping the waterers clean is essential.

Semi-Annual Maintenance Tasks

Deep Cleaning the Coop
Deep cleaning the coop is recommended at least twice a year. This involves washing down all the surfaces in the coop. A mixture of bleach and water can be used to disinfect and sanitize the coop completely. During the deep cleaning process, you can also try sprinkling some diatomaceous earth in the coop. It has been found to help in getting rid of parasites such as lice and mites.

Winter-Proof the Coop
Winter months can be stressful for chickens, and it is important to prepare the coop before winter. In colder months, you may notice that your layers stop laying eggs or lay fewer eggs than they normally would. This is because they do not receive enough daylight to stimulate egg production. Hens typically require a minimum of 16 hours of daylight for egg production. During winter months, egg production will inevitably decrease due to a lack of sufficient daylight. To avoid this situation, put a source of artificial light in the coop during the winter months. This will help to keep your layers productive.

It is also recommended that you add more layers of bedding as winter approaches. Deep litter will help to keep the coop well insulated during the cold season. You may also need heaters for your waterers to keep them from freezing up when temperatures get low.

Keeping Your Flock Healthy

When it comes to health management for your backyard flock, health management will fall into three basic categories:
1. Prevention of disease

2. Early intervention

3. Treatment of disease

Prevention of Disease

The best thing to do for your chickens is not to let them get sick at all. Of course, in some circumstances, this is not always within your control, but in most cases, you can take measures to reduce the risk of diseases. These preventive measures include:

a) Making sure your chicks are vaccinated against common poultry diseases.

b) If your chicks are not vaccinated, using medicated feed may help in boosting the immune system.

c) Maintaining a clean, well-aerated living environment will help to minimize the risks of infections.

d) Providing your chickens with well-balanced, age-appropriate feed to ensure that their basic nutritional needs are being met.

e) Ensuring that your flock has access to clean drinking water at all times.

f) Protecting your flock from extreme conditions such as extreme heat or cold.

g) Keeping your flock safe from predators.

Early Intervention

If you catch signs of poor health early, chances are that the disease will be much easier to treat. This will also prevent the disease from spreading to the entire flock. For this to happen, you need to spend time regularly observing your chickens and taking note of any abnormal behavior.

Here are some warning signs that point to possible underlying conditions that you need to address.

a) Discharge from the nostrils or eyes

b) Droopy wings

c) Lethargy and poor movement/coordination

d) Poor appetite

e) A hen that suddenly stops laying eggs for no apparent reason

f) Weight loss or retarded growth

g) Ruffled feathers

h) Inability to hold their head up

i) Wounds on legs

j) Loss of feathers

When you notice your chicken has any of the above symptoms, it is best to seek help from a vet as soon as possible. You can separate the ill chicken from the rest of the flock to avoid having a disease spread to the rest of the flock.

Treatment of Disease

Getting treatment for any sick chickens is important if you do not want to lose your birds to diseases. Have a vet check on any sick birds and advise on the recommended treatment or next step. Chicken diseases spread pretty fast, and one infected chicken can easily wipe out your entire flock if it is not treated in time.

Common Chicken Diseases

FowlPox

Fowlpox is a common poultry disease. It is spread by mosquitoes, though it also spreads from one chicken to another. Although fowlpox is not necessarily fatal, it can cause death in weak and younger chickens. Fowlpox will usually infect birds for 10-14 days. Some of the symptoms of fowl pox include:

- Comb sores
- White spots on the skin
- Egg production ceases
- Mouth ulcers

Chickens can be vaccinated against fowl pox to minimize the risk of contraction. However, once the birds have contracted the disease, treatment is usually done using supplements of Vitamins A, D, and E. Sick chickens should be fed soft food only until they heal.

Botulism

This disease is caused by food or water contamination. While botulism is not infectious, if your chickens share the same feeder and waterer, they can all get the disease from the contaminated water or feed. Some of the common symptoms of botulism include:

- Feather loss
- Weakness
- Tremors and shakes
- Paralysis which eventually leads to death

If the disease is treated early, the bird can be saved. Some people use a teaspoon of Epsom salts in warm water as a home remedy.

Infectious Bronchitis

This is one of the most common illnesses in backyard flocks. This disease can easily wipe out an entire flock if left untreated. Here are some of the symptoms to be on the lookout for:

- Loss of appetite
- Decreased egg production
- Lethargy
- Eye and nasal discharge
- Misshapen eggs

Ultimately with good care and maintenance, your chickens can live happy and productive lives. Taking care of your pet is not just fulfilling, but it also ensures that you get good quality eggs from your backyard chickens.

Like any other venture, you will learn more and more about the best way to meet your chickens' needs with experience. With time it will be easier for you to identify any problems in the flock and adjust accordingly. Ultimately to raise healthy backyard chickens, you do not need to have a lot of pastureland or spend a lot of money. You can still keep things simple and as natural as possible and raise a productive, happy, and healthy flock.

Conclusion

Nurturing a living thing is probably one of the most rewarding things anyone can ever do. The fulfillment and joy that come from seeing something thrive under your care are invaluable. This is why taking the time to understand how best to take care of your chickens is not just good for your pets but also good for you. Taking the chance to raise your own backyard flock will be so much easier now that you know how to go about it.

Whether you are interested in raising chickens for eggs, for meat, or just the simple pleasure of having an easy-going pet, there are plenty of benefits that come with raising chickens in your backyard. As long as you are willing to dedicate a little time and energy to the care of your chickens, the rewards are bound to outweigh any challenges that you may encounter in the process.

The important thing to remember is that you do not need dozens of chickens to get started. A simple flock of six birds, if well taken care of, can supply you with enough eggs for your family and even surplus that can be sold. Start small and build your flock slowly as you get more knowledgeable about raising chickens, caring for them, and keeping them healthy.

One of the best things about raising chickens is that it is relatively inexpensive. Most of what you need to raise and care for chicken are things that can be easily improvised and made at home. This means that costs should not come between you and the dream of having a flock of healthy backyard chickens to call your own. With a little capital, you will be able to get back most of what you need to get started.

Since you have already taken the first step by equipping yourself with the information and knowledge you need, the next step is simply to start using the knowledge you have acquired and get started setting up for your chickens. The information in this book is timeless and will come in handy whether you choose to start raising chickens today, or sometime in the future.

We hope that you know we have all the tools and information you need to pursue this fulfilling hobby. Finally, if you found the content in this book useful, a review on Amazon is always appreciated.

Here's another book by Dion Rosser
that you might be interested in

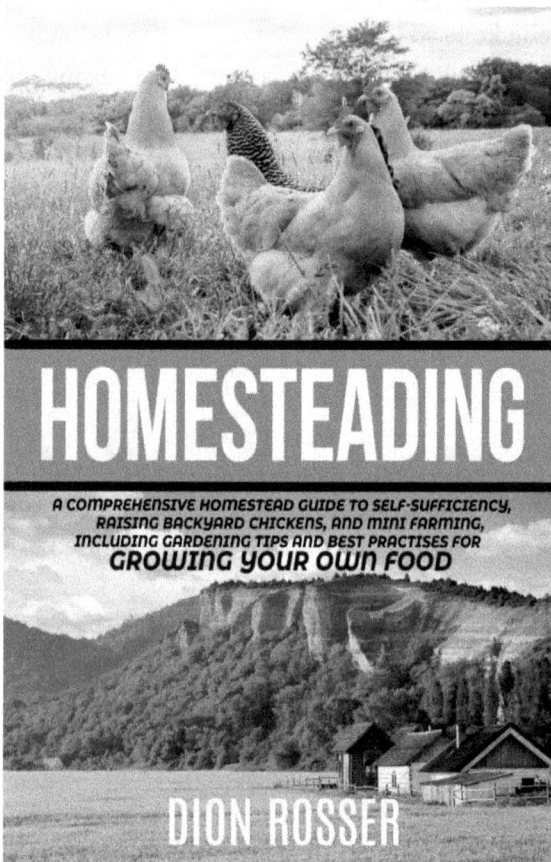

HOMESTEADING

A COMPREHENSIVE HOMESTEAD GUIDE TO SELF-SUFFICIENCY,
RAISING BACKYARD CHICKENS, AND MINI FARMING,
INCLUDING GARDENING TIPS AND BEST PRACTISES FOR
GROWING YOUR OWN FOOD

DION ROSSER